$$A = \frac{a^2 x}{3}$$

How can you
get this kind
of thing?

$$B = \frac{b^2(x+h)}{3}$$

Get it the
same way

$$\frac{x}{x+h} = \frac{a}{b}$$

Plan completed!
work from data
to unknown

MATHEMATICAL DISCOVERY

GEORGE POLYA

PROFESSOR EMERITUS OF MATHEMATICS, STANFORD UNIVERSITY

New York · London · Sydney   JOHN WILEY & SONS, INC.

MATHEMATICAL

On understanding, learning, and teaching problem solving

DISCOVERY VOLUME II

10 9 8 7 6 5 4

Copyright © 1965 by John Wiley & Sons, Inc.

*All Rights Reserved. This book or any part
thereof must not be reproduced in any form without
the written permission of the publisher.*

LIBRARY OF CONGRESS CATALOG CARD NUMBER: 62-8784
PRINTED IN THE UNITED STATES OF AMERICA

ISBN 0 471 69335 9

PREFACE TO THE CORRECTED PRINTING

The text of the first printing is reprinted here with a few minor changes. An appendix is added which contains 35 problems with solutions. These problems supplement various chapters in both volumes of the work; their numbering indicates where they should be inserted between the problems of the first printing.

GEORGE POLYA

Stanford University
June 1967

PREFACE

The present second volume follows the plan, and attempts to carry into effect the intentions, indicated in the preface to the first volume. The index at the end of this volume refers to both volumes; moreover it contains references to selected parallel passages of my related books, which may be of service, I hope, to interested readers.

The Committee on the Undergraduate Program in Mathematics supported the preparation of the manuscript of this second volume by funds granted by the National Science Foundation. I wish to express my thanks to the Committee for their support and encouragement. I wish to thank also the Publisher for courteous help and the careful printing.

The last chapter of this book was dedicated to Charles Loewner on the occasion of his seventieth birthday; I wish to reiterate here the expression of my high esteem and of my feelings of friendship.

GEORGE POLYA

Zurich, Switzerland
October 1964

FROM THE PREFACE TO VOLUME I

A method of solution is perfect if we can foresee
from the start, and even prove, that following
that method we shall attain our aim.
LEIBNITZ: *Opuscules,* p. 161.

1. Solving a problem means finding a way out of a difficulty, a way around an obstacle, attaining an aim which was not immediately attainable. Solving problems is the specific achievement of intelligence, and intelligence is the specific gift of mankind: solving problems can be regarded as the most characteristically human activity. The aim of this work is to understand this activity, to propose means to teach it, and, eventually, to improve the problem-solving ability of the reader.

2. This work consists of two parts; let me characterize briefly the role of these two parts.

Solving problems is a practical art, like swimming, or skiing, or playing the piano: you can learn it only by imitation and practice. This book cannot offer you a magic key that opens all the doors and solves all the problems, but it offers you good examples for imitation and many opportunities for practice: if you wish to learn swimming you have to go into the water, and if you wish to become a problem solver you have to solve problems.

If you wish to derive the most profit from your effort, look out for such features of the problem at hand as may be useful in handling the problems to come. A solution that you have obtained by your own effort or one that you have read or heard, but have followed with real interest and insight, may become a *pattern* for you, a model that you can imitate with advantage in solving similar problems. The aim of Part One is to familiarize you with a few useful patterns.

It may be easy to imitate the solution of a problem when solving a closely similar problem; such imitation may be more difficult or scarcely possible if the similarity is not so close. Yet there is a deep-seated human desire for more: for some device, free of limitations, that could

ix

solve all problems.   This desire may remain obscure in many of us, but it becomes manifest in a few fairy tales and in the writings of a few philosophers.   You may remember the tale about the magic word that opens all the doors.   Descartes meditated upon a universal method for solving all problems, and Leibnitz very clearly formulated the idea of a perfect method.   Yet the quest for a universal perfect method has no more succeeded than did the quest for the philosopher's stone which was supposed to change base metals into gold; there are great dreams that must remain dreams.   Nevertheless, such unattainable ideals may influence people: nobody has attained the North Star, but many have found the right way by looking at it.   This book cannot offer you (and no book will ever be able to offer you) a universal perfect method for solving problems, but even a few small steps toward that unattainable ideal may clarify your mind and improve your problem-solving ability. Part Two outlines some such steps.

3.I wish to call *heuristics* the study that the present work attempts, the study of means and methods of problem solving.   The term heuristic, which was used by some philosophers in the past, is half-forgotten and half-discredited nowadays, but I am not afraid to use it.

In fact, most of the time the present work offers a down-to-earth practical aspect of heuristic: I am trying, by all the means at my disposal, to entice the reader to do problems and to think about the means and methods he uses in doing them.

In most of the following chapters, the greater part of the text is devoted to the broad presentation of the solution of a few problems.   The presentation may appear too broad to a mathematician who is not interested in methodical points.   In fact, what is presented here are not merely solutions but *case histories* of solutions.   Such a case history describes the sequence of essential steps by which the solution has been eventually discovered, and tries to disclose the motives and attitudes prompting these steps.   The aim of such a careful description of a particular case is to suggest some general advice, or pattern, which may guide the reader in similar situations.   The explicit formulation of such advice or such a pattern is usually reserved for a separate section, although tentative first formulations may be interspersed between the incidents of the case history.

Each chapter is followed by examples and comments.   The reader who does the examples has an opportunity to apply, clarify, and amplify the methodical remarks offered in the text of the chapter.   The comments interspersed between the examples give extensions, more technical or more subtle points, or incidental remarks.

How far I have succeeded I cannot know, but I have certainly tried hard to enlist the reader's participation. I have tried to fix on the printed page whatever modes of oral presentation I found most effective in my classes. By the case histories, I have tried to familiarize the reader with the atmosphere of research. By the choice, formulation, and disposition of the proposed problems (formulation and disposition are much more important and cost me much more labor than the uninitiated could imagine) I have tried to challenge the reader, awake his curiosity and initiative, and give him ample opportunity to face a variety of research situations.

4. This book deals most of the time with mathematical problems. Nonmathematical problems are rarely mentioned, but they are always present in the background. In fact, I have carefully taken them into consideration and have tried to treat mathematical problems in a way that sheds light on the treatment of nonmathematical problems whenever possible.

This book deals most of the time with elementary mathematical problems. More advanced mathematical problems, however, although seldom referred to, led me to the conception of the material included. In fact, my main source was my own research, and my treatment of many an elementary problem mirrors my experience with advanced problems which could not be included in this book.

5. This book combines its theoretical aim, the study of heuristics, with a concrete, urgent, practical aim: to improve the preparation of high school mathematics teachers.

I have had excellent opportunity to make observations and form opinions on the preparation of high school mathematics teachers, for all my classes have been devoted to such teachers in the last few years. I hope to be a comparatively unprejudiced observer, and as such I can have but one opinion: *the preparation of high school mathematics teachers is insufficient.* Furthermore, I think that all responsible organizations must share the blame, and that especially both the schools of education and the departments of mathematics in the colleges should very carefully revise their offerings to teachers if they wish to improve the present situation.

What courses should the colleges offer to prospective high school teachers? We cannot reasonably answer this question, unless we first answer the related question: *What should the high schools offer to their students?*

Yet this question is of little help, you may think, because it is too controversial; it seems impossible to give an answer that would command sufficient consensus. This is unfortunately so; but there is an aspect of this question about which at least the experts may agree.

Our knowledge about any subject consists of *information* and of *know-how*. If you have genuine *bona fide* experience of mathematical work on any level, elementary or advanced, there will be no doubt in your mind that, in mathematics, know-how is much more important than mere possession of information. Therefore, in the high school, as on any other level, we should impart, along with a certain amount of information, a certain degree of *know-how* to the student.

What is know-how in mathematics? The ability to solve problems— not merely routine problems but problems requiring some degree of independence, judgment, originality, creativity. Therefore, the first and foremost duty of the high school in teaching mathematics is to emphasize *methodical work in problem solving*. This is my conviction; you may not go along with it all the way, but I assume that you agree that problem solving deserves some emphasis—and this will do for the present.

The teacher should know what he is supposed to teach. He should show his students how to solve problems—but if he does not know, how can he show them? The teacher should develop his students' know-how, their ability to reason; he should recognize and encourage creative thinking—but the curriculum he went through paid insufficient attention to his mastery of the subject matter and no attention at all to his know-how, to his ability to reason, to his ability to solve problems, to his creative thinking. Here is, in my opinion, the worst gap in the present preparation of high school mathematics teachers.

To fill this gap, the teachers' curriculum should make room for *creative work on an appropriate level.* I attempted to give opportunity for such work by conducting seminars in problem solving. The present work contains the material I collected for my seminars and directions to use it; see the "Hints to Teachers, and to Teachers of Teachers" at the end of this volume, pp. 209–212. This will, I hope, help to improve the mathematics teacher's preparation; at any rate, this is the practical aim of the present work.

I believe that constant attention to both aims mentioned, the theoretical and the practical, made me write a better book. I believe too that there is no conflict between the interests of the various prospective readers (some concerned with problem solving in general, others with improving their own ability, and still others with improving the ability of their students). What matters to one type of reader has a good chance to be of consequence to the others.

6. The present work is the continuation of two earlier ones, *How to Solve It* and *Mathematics and Plausible Reasoning;* the two volumes of the latter have separate titles: *Induction and Analogy in Mathematics*

(vol. 1) and *Patterns of Plausible Inference* (vol. 2). These books complete each other without essential overlapping. A topic considered in one may be reconsidered in another, but then the treatment is different: other examples, other details, or other aspects are offered. And so it does not matter much which one is read first and which one is read later.

For the convenience of the reader, the three works will be compared and corresponding passages listed in a cumulative index at the end of the second volume of this book, *Mathematical Discovery*.

7. To publish the first part of a book when the second part is not yet available entails certain risks. (There is a German proverb; "Don't show a half-built house to a fool.") These risks are not negligible; yet, in the interest of the practical aim of this work, I decided not to delay the publication of this volume; see p. 210.

This first volume contains Part One of the work, Patterns, and two chapters of Part Two, Toward a General Method.

The four chapters of Part One have more extensive collections of problems than the later chapters. In fact, Part One is in many ways similar to a collection of problems in analysis by G. Szegö and the author (see the Bibliography). There are, however, obvious differences: in the present volume the problems proposed are much more elementary, and methodical points are not only suggested but explicitly formulated and discussed.

The second chapter of Part Two is inspired by a recent work of Werner Hartkopf (see the Bibliography). I present here only some points of Hartkopf's work which seem to me the most engaging, and I present them as they best fit my conception of heuristics, with suitable examples and additional remarks.

G.P.

HINTS TO THE READER

Section 5 of chapter 2 is quoted as sect. 2.5, subsection (3) of section 5 of chapter 2 as sect. 2.5(3), example 61 of chapter 3 as ex. 3.61.

HSI and MPR are abbreviations for titles of books by the author which will be frequently quoted; see the Bibliography.

Iff. The abbreviation "iff" stands for the phrase "if and only if."

†. The sign † is prefixed to examples, comments, sections, or shorter passages that require more than elementary mathematical knowledge (see the next paragraph). This sign, however, is not used when such a passage is very short.

Most of the material in this book requires only *elementary mathematical knowledge,* that is, as much geometry, algebra, "graphing" (use of coordinates), and (sometimes) trigonometry as is (or ought to be) taught in a good high school.

The problems proposed in this book seldom require knowledge beyond the high school level, but, with respect to difficulty, they are often a little above the high school level. The solution is fully (although concisely) presented for some problems, only a few steps of the solution are indicated for other problems, and sometimes only the result is given.

Hints that may facilitate the solution are added to some problems (in parentheses). The surrounding problems may provide hints. Especial attention should be paid to the introductory lines prefixed to the examples (or to certain groups of examples) in some chapters.

The reader who has spent serious effort on a problem may benefit from the effort even if he does not succeed in solving the problem. For example, he may look at some part of the solution, try to extract some helpful information, and then put the book aside and try to work out the rest of the solution by himself.

The best time to think about methods may be when the reader has finished solving a problem, or reading its solution, or reading a case history. With his task accomplished and his experience still fresh in mind, the reader, in *looking back* at his effort, can profitably explore the nature of the difficulty he has just overcome. He may ask himself many useful questions: "What was the decisive point? What was the main difficulty? What could I have done better? I failed to see this point: which item of knowledge, which attitude of mind should I have had to see it? Is there some trick worth learning, one that I could use the next time in a similar situation?" All these questions are good, and there are many others—but the best question is the one that comes spontaneously to mind.

CONTENTS FOR VOLUME I

PART TWO
TOWARD A GENERAL METHOD, 115

CONTENTS FOR VOLUME II

TOWARD A GENERAL METHOD

*Human wisdom remains always one and the same
although applied to the most diverse objects and it is
no more changed by their diversity than the sunshine
is changed by the variety of objects
which it illuminates.*

DESCARTES: Rule I, *Œuvres,* vol. X, p. 360.

CHAPTER 7

GEOMETRIC REPRESENTATION OF THE PROGRESS OF THE SOLUTION

It is very helpful to represent these things in this fashion since nothing enters the mind more readily than geometric figures.
DESCARTES: *Œuvres,* vol. X, p. 413; Rules for the Direction of the Mind, Rule XII.

7.1. Metaphors.

It happened about fifty years ago when I was a student; I had to explain an elementary problem of solid geometry to a boy whom I was preparing for an examination, but I lost the thread and got stuck. I could have kicked myself that I failed in such a simple task, and sat down the next evening to work through the solution so thoroughly that I shall never again forget it. Trying to see intuitively the natural progress of the solution and the concatenation of the essential ideas involved, I arrived eventually at a geometric representation of the problem-solving process. This was my first discovery, and the beginning of my lifelong interest, in problem solving.

I was guided to the geometric image that finally emerged by a group of usual metaphorical expressions. That the language is full of metaphors (dead, half-dead, and live) has been observed often enough. I do not know whether it has been also observed that many of these metaphors are interdependent: they are connected, they are associated, they form clusters, more or less loose, more or less overlapping families. At any rate, there is an extensive family of metaphorical expressions which have two things in common: they are all concerned with that basic human activity of problem solving and they are all suggesting the same geometric configurations.

1

Discovering the solution is finding a *connection* between formerly separated things or ideas (the things we have and the things we want, the data and the unknown, the hypothesis and the conclusion). The farther apart the things connected stood originally, the more credit is due to the discoverer for connecting them. We sometimes see the connection under the guise of a *bridge:* a great discovery strikes us as bridging over a deep chasm between two widely separated ideas. We often see the connection realized by a *chain:* a proof appears as a *concatenation* of arguments, as a chain, perhaps a long chain, of conclusions. The chain is not stronger than its weakest *link,* and there is no valid proof, no uninterrupted chain of reasoning, if even one link is missing. More often still we use a *thread* for mental connecting. We have all listened to the professor who lost the thread, or got himself entangled in the threads, of his argument. He was obliged to peek in his notes before resuming the thread of his lecture, and we were all tired when he gathered up the threads to a final conclusion. A slender thread becomes a mere geometric line, the things connected become mere geometric points, and there emerges, almost unavoidably, a diagrammatic, graphic image of the concatenation of mathematical conclusions.

Let us look now at some geometric figures instead of merely listening to figures of speech.

7.2. What is the problem?

We need an example, and I choose a very simple problem of solid geometry:[1]

Find the volume F of the frustum of a right pyramid with square base. Given the altitude h of the frustum, the length a of a side of its upper base, and the length b of a side of its lower base.

(A pyramid with square base is a *right pyramid* if its height meets the base at the center. The *frustum* of a pyramid is the portion of the pyramid between its base and a plane parallel to its base; this parallel plane contains a face of the frustum which we call the *upper base* of the frustum; its *lower base* is the base of the original full pyramid, its *altitude* is the perpendicular distance between its bases.)

The first step in solving the problem is to concentrate on its aim. *What do you want?* we ask ourselves and we represent as sharply as we can the shape of which we wish to find the volume F (see the left side of Fig. 7.1). The mental situation is appropriately represented by a single point, labeled F, on which our whole attention should be focused (see the right-hand side of Fig. 7.1).

---

[1] Very similar to, but even simpler than, the problem originally considered. See the Bibliography, papers 1 and 3 of the author.

F
o

What do you want?

Fig. 7.1. Concentrate on one point: the aim.

Yet we cannot find the unknown F if nothing is given. *What are the data?* we ask ourselves, or *What have you?* and our attention emphasizes those lines of the figure of which the length is given, a, b, and h, see the left-hand side of Fig. 7.2. (The square with side a is above, the square with side b below, the solid considered.) To portray the changed mental situation, three new points emerge, labeled a, h, and b, representing the data and separated from the unknown F by a gap, an *open space* in Fig. 7.2. This open space symbolizes the open question; our problem aims at connecting the unknown F with the data a, h, and b, we have to bridge the gap.

7.3. That's an idea!

We start working at our problem by visualizing its aim, its unknown, its data. This initial phase of our work is fittingly pictured by Figs.7.1 and 7.2. Yet how should we proceed from here, which course should we adopt?

If you cannot solve the proposed problem, look around for an appropriate related problem.

In our case, we do not have to look far. In fact, *what is the unknown?* The volume of the frustum of a pyramid. And what is such a frustum? How is it *defined?* As a portion of a full pyramid. Which portion? The portion between—No, enough of that; let us *restate it differently:*

F o

What have you?

o   o   o
a h b

Fig. 7.2. The open question: a gap to bridge over.

The frustum is the portion that remains when we cut off from the full pyramid a smaller pyramid by a plane parallel to the base. In our case, see Fig. 7.3, the base of the big (full) pyramid is a square of area b^2. *If* we knew the volumes of these two pyramids, say B and A, respectively, we could find the volume of the frustum:

$$F = B - A$$

Let us find these volumes B and A. That's an idea!

Thus we have reduced our original problem, to find F, to two appropriately related auxiliary problems, to find A and B. To express this reduction graphically, we introduce two new points, labeled A and B, into the space between the data a, h, b and the unknown F. We join both A and B to F with slanting lines and thereby indicate the essential relation between these three quantities: starting from A and B we can reach F; the solution of the problem about F can be based on the solution of the two problems about A and B.

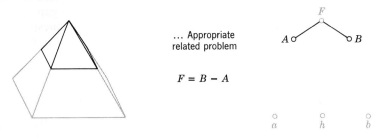

... Appropriate related problem

$F = B - A$

Fig. 7.3. If you cannot solve the proposed problem, look around for an . . . .

Our work is not yet finished; we have two new unknowns to find, A and B; there are two pendent points separated from the data by a gap in Fig. 7.3. The situation looks hopeful, however; the pyramid is a more familiar figure than the frustum, and although we have two unknowns, A and B, to find instead of one, F, the two are of similar nature and similarly related to the data a and b, respectively. Correspondingly, the graphical representation of the mental situation in Fig. 7.3 is symmetrical. The line FA is inclined toward the given a, FB toward the given b. We have started bridging the open space between the original unknown and the data; the remaining gap is narrower.

7.4. Developing the idea

Where are we now? *What do you want?* We want to find the unknowns A and B. *What is the unknown A?* The volume of a pyramid.

How can you get this kind of thing? How can you find this kind of unknown? From what data can you derive this kind of unknown? The volume of the pyramid can be computed if we have two data, the area of the base and the height of the pyramid; the volume is, in fact, the product of these two quantities divided by 3. The height is not given, but we can still consider it. Let us call it *x*. Then

$$A = \frac{a^2 x}{3}$$

On the left-hand side of Fig. 7.4 the small pyramid above the frustum appears in more detail; its height *x* is emphasized. The present stage of our work is graphically represented on the right-hand side of Fig. 7.4; a new point *x* appears above the data, and slanting lines join *A* to *x* and *a*, indicating that *A* can be reached from *x* and *a*, that *A* can be expressed in terms of *x* and *a*. Although there are still two unknowns to find (there are still loose ends dangling in mid air in Fig. 7.4) progress

Fig. 7.4. A first connection with the data, but loose ends dangling in midair.

has been made; we have succeeded in connecting the unknown *F* at least with one of the data, with *a*.

At any rate the next step is obvious. The unknowns *A* and *B* are of similar nature (they are symmetrically represented in Fig. 7.3); we have expressed the volume *A* in terms of base and height, and we can express the volume *B* analogously:

$$B = \frac{b^2 (x + h)}{3}$$

In the left-hand portion of Fig. 7.5, the big pyramid containing the frustum appears in more detail, its height *x + h* is emphasized. In the right-hand portion of Fig. 7.5 three new slanting lines appear, joining *B* to *b*, *h*, and *x*. These lines indicate that *B* can be attained from *b*, *h*, and *x*, that *B* can be expressed in terms of *b*, *h*, and *x*. And so just one

Get it the same way

$$B = \frac{b^2(x + h)}{3}$$

Fig. 7.5. Just one question remains pending.

pending point—point x—remains unconnected with the data. The gap is narrowed down; it is now between x and the data.

What is the remaining unknown? It is x, the length of a line. *How can you find this kind of unknown?* How can you get this kind of thing?

The most usual thing is to obtain the length of a line from a triangle—from a right triangle, if possible—or from a pair of similar triangles. Yet there is no usable triangle in the figure, and there should be one with a side x. Such a triangle would lie in a plane passing through the altitude of the small pyramid with volume A; this plane would also pass through the altitude of the big pyramid with volume B, which is similar to the small one. Yes, similar triangles in a plane passing through the altitude—and *parallel to a given side* of the base of one of these pyramids! That's it! That finishes it!

A pair of similar triangles appears in Fig. 7.6 from which x can be conveniently computed by the proportion

$$\frac{x}{x + h} = \frac{a}{b}$$

But details are not important at this stage; what is all important now is that x can be expressed in terms of the three data a, h, and b. The three new slanting lines emerging in the right-hand portion of Fig. 7.6 indicate just that by joining x to a, h, and b.

And how can you get *this* kind of thing?

$$\frac{x}{x + h} = \frac{a}{b}$$

Fig. 7.6. We have succeeded in bridging the gap.

Done! We have succeeded in bridging the gap, in establishing an uninterrupted connection between the unknown F and the data a, h, and b through the intermediaries (auxiliary unknowns) A, B, and x.

7.5. Carrying it out

Is the problem solved? Not yet—not quite. We are required to express the volume F of the frustum in terms of the data, a, h, and b, and this is not yet done. Yet the more important and more exciting first part of our work is behind us; the remaining task is a much calmer affair, much more straightforward.

There was an element of adventure in the first part of our work. At each stage, we *hoped* that the next step would bring us nearer to our goal, to bridging the gap. Yes, we hoped, but we were not certain; at each stage we had to *invent,* and risk, the next step. But now no more invention or risk is needed; we foresee that we can safely reach the unknown F from the data a, h, and b just by following the threads of the uninterrupted connection represented in Fig. 7.6.

We begin the second part of our work where we have ended the first part. We tackle first the auxiliary unknown x that we have introduced last; from the last equation of sect. 7.4 we obtain

$$x = \frac{ah}{b - a}$$

Then we substitute this value for x in the two foregoing equations of sect. 7.4, obtaining

$$A = \frac{a^3 h}{3(b - a)}, \qquad B = \frac{b^3 h}{3(b - a)}$$

(The analogy between these two results is comforting.) We use last the equation that we obtained first, in sect. 7.3:

$$F = B - A = \frac{b^3 - a^3}{b - a} \frac{h}{3}$$

$$F = \frac{a^2 + ab + b^2}{3} h$$

This is the desired expression.

The work of the present section is aptly symbolized by Fig. 7.7 where each connecting line carries an arrow indicating the direction in which we have just used the connection. We started from the data a, h, and b, and proceeded hence through the intermediate auxiliary unknowns x, A, and B to our original, principal unknown F, expressing these quantities one after the other in terms of the data.

a h b **Fig. 7.7.** Working from the data to the unknown.

7.6. A slow-motion picture

Figures 7.1 to 7.7 show successive stages of the solution; we assemble these seven figures into one composite picture, Fig. 7.8. (Figure 7.7 is merged into Fig. 7.6. Fig. 7.8 is reproduced in black and red on the endpapers immediately inside the cover at the beginning and the end of the book; the details on which the attention is focused are emphasized by red—red meaning, of course, "burning interest.") View the sequence in Fig. 7.8 from left to right. Viewing it in fast succession, we have a kind of cinematographic representation of the problem-solvers progress, of the march of discovery. Or, viewing Fig. 7.8 at low speed, we have a sort of slow-motion picture of the evolving solution which leaves us time for careful observation of significant details.

In Fig. 7.8, each stage of the solution (each mental situation of the problem-solver) is represented at several levels; the items pertaining to the same situation are vertically arranged, one under the other. Thus the solution progresses along four parallel lines, on four different levels.

On the uppermost level, the *image level,* we see the evolution of the investigated geometric figure in the problem solver's mind. At each stage, the problem solver has a mental picture of the geometric figure he explores, but this picture changes in transition to the next stage; some details may recede into the background, other details come to our attention, new details are added.

Going down to the next layer, we attain the *relational* level. In graphical representation, the objects considered (unknown, data, auxiliary unknowns) are symbolized as points, and the relations connecting the objects are indicated by lines connecting the symbolizing points.

Immediately below the relational level we find the *mathematical* level, which consists of formulas and which contrasts with the relational level. On the relational level, the system of all relations obtained up to the moment in question is exhibited; the last obtained relation is emphasized by color (it is in the focus of attention) but it is not exhibited with more details than the preceding. On the mathematical

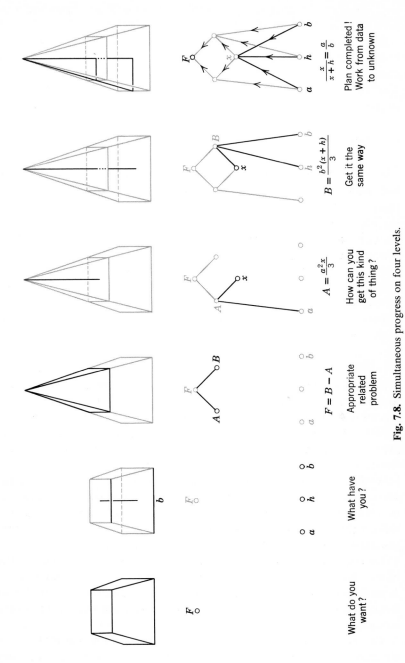

Fig. 7.8. Simultaneous progress on four levels.

9

level, the last obtained relation is fully shown and the preceding relations are not shown at all.

Going down still further we attain what is for us the basic layer, the *heuristic* level. At each stage, it exhibits a simple, natural (generally applicable!) question or suggestion which has enabled us to reach that stage. Our basic concern is to study the nature of such questions and suggestions.

7.7. A preview

We wish to view and review Fig. 7.8; we wish to compare it and correlate it with our past experience. Thus behind the story this figure tells, we wish to perceive points of general interest which deserve further study. In the graphical narrative of the solution of a single problem, we shall find useful preindications about the general questions treated in the following chapters. We shall survey these chapters one by one, in due order.

(Each of the following sections of the present chapter previews a following chapter and has the same title and a corresponding number as the chapter that is previews.)

And now let us try to dig down through the particular story to the underlying general ideas.

7.8. Plans and programs

In viewing the stages of Fig. 7.8 in sequence, we see how the problem solver's attention ranges over the geometric figure he explores, how he takes possession of more and more details of this figure, and how he builds up step by step the system of connections which constitutes the plan of his solution. In viewing the unfolding of the solution attentively, we may discern in it several phases and activities.

We have already observed (in sect. 7.5) the contrast between the two portions of the solution: in the first portion (illustrated by Figs. 7.1 to 7.6) we work downward, from the unknown to the data, in the second portion (illustrated by the single Fig. 7.7) we work upward, from the data to the unknown.

Yet even in the first portion we may distinguish two phases. In the initial phase, Figs. 7.1 and 7.2, the problem solver's main effort is aimed at understanding his problem. In the latter phase, Figs. 7.3 to 7.6, he develops the system of logical connections, he constructs a *plan* of the solution.

This latter phase, the devising of a plan, seems to be the most essential part of the problem solver's work; we shall examine it more closely in chapter 8.

7.9. Problems within problems

In surveying Fig. 7.8, we may observe that the problem solver, in working at his original (proposed, main) problem, encountered several *auxiliary problems* ("helping" problems, subproblems). In seeking the volume of the frustum, he was led to seek the volume of a full pyramid, then that of another full pyramid, then the length of a line. To arrive at his original unknown F, he had to pass through the auxiliary unknowns A, B, and x. Just a little experience in solving mathematical problems may be enough to convince us that such breaking up of the proposed problem into several subproblems is typical (see sect. 2.5(3), for instance).

We shall thoroughly examine the role of auxiliary problems and distinguish various kinds of auxiliary problems.

7.10. The coming of the idea

Which of the various steps of the solution illustrated by Fig. 7.8 is the most important? The popping up of the full pyramid (Fig. 7.3) I think—and I think too that most people who have had some experience in these matters and have devoted some thought to their experience will agree. Introducing the full pyramid and conceiving the frustum as the difference of two full pyramids is the *decisive idea* of the solution; for most problem solvers the rest of the solution will be easier, more obvious, more routine—it may be almost completely routine for the more experienced problem solvers.

The emergence of the decisive idea is not very impressive in the present case, but we should not forget that the problem we are discussing is very simple. Conceiving the decisive idea, seeing a sudden light after a long period of tension and hesitation may be very impressive; it may be a great experience which the reader should not miss.

7.11. The working of the mind

In the graphical representation of Fig. 7.8, the most conspicuous sign of progress is the introduction of more and more detail. As the problem solver advances, more and more lines appear both in the geometric figure and in the relational diagram. Behind the increasing complexity of the figure we should perceive a growing structure in the problem solver's mind. At each significant step he brings in some relevant knowledge; he recognizes some familiar configuration, he applies some known theorem. Thus the work of the problem solver's mind appears as recalling relevant elements of his experience and connecting them with the problem at hand, a work of *mobilization* and *organization*.

We have already had the opportunity in the foregoing to make such remarks (especially in ex. 2.74), and we shall have the opportunity to explore this point as well as other aspects of the working of the mind.

7.12. The discipline of the mind

Figure 7.8, which represents the progress of the solution on four different levels, gives some indication about the problem solver's work. We certainly want to know how he is working, but we are even more anxious to know how he ought to be working. Can Fig. 7.8 give some indication of this?

On the lowest level of Fig. 7.8 there is a sequence of questions and suggestions motivating the successive steps of the problem solver's work. These questions and suggestions are simple, natural, and very general; they have guided the problem solver to the solution of the simple problem we have chosen as example, and they may guide us in innumerable other cases. If there is a discipline of the mind (some system of guidelines, some body of maxims or rules in the direction of the universal method sought by Descartes and Leibnitz) there is a good chance that the questions and suggestions aligned in the basic layer of Fig. 7.8 somehow belong to it. We must thoroughly examine this point.

Examples and Comments on Chapter 7

7.1. *Another approach* to the problem stated in sect. 7.2. The base of the frustum is in a horizontal plane (on the desk). By four vertical planes passing through the four sides of the top of the frustum (see Fig. 7.9) we divide it into 9 polyhedrons:

> a prism with square base of volume Q,
> four prisms with triangular base of volume T each,
> four pyramids with square base of volume P each.

Compute F, following the approach indicated by Fig. 7.10.

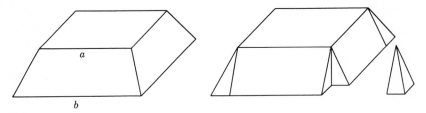

Fig. 7.9. By four vertical planes.

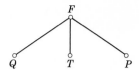

Fig. 7.10. Another approach.

7.2. Express the two solutions of the same problem given in sect. 2.5(3) and 2.5(4) diagrammatically (the quantities involved by points, the connecting relations by connecting lines).

7.3. *The search for a proof.* Proposition 4 in the Eleventh Book of Euclid's Elements can be stated as follows:

If a straight line passes through the point of intersection of two given straight lines and is perpendicular to both, then it is also perpendicular to any third straight line that lies in the plane of the two given lines and passes through their point of intersection.

We wish to analyze a proof of this proposition, visualize its structure, and understand the motives of its discovery in using the geometric representation of the problem solver's progress developed in this chapter. In relying on the analogy with the discussion illustrated by Figs. 7.1 to 7.8, we shall deal with the present case somewhat more concisely.

We are here principally concerned with the formation of the proof in the problem solver's mind. But the proposition to be proved is interesting too; it states a basic fact of solid geometry. Even the logical form of the proposition is interesting. The teacher who said "Two bad boys spoil the whole class" was possibly wrong, but his statement has the same form as Euclid's proposition we are about to prove.

(1) *Working backward.* We draw Fig. 7.11, introduce suitable notation, and then we restate the proposition we intend to prove in standard form, split into hypothesis and conclusion:

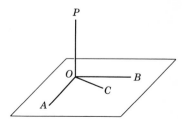

Fig. 7.11. Two bad boys spoil the whole class.

Hypothesis: The three lines OA, OB, and OC pass through the same point O and are in the same plane, but they are different; and

$$PO \perp OA \qquad PO \perp OB$$

Conclusion: Then also

$$PO \perp OC$$

"What is the conclusion?"

That the line PO is perpendicular to OC. Or that $\angle POC$ is a right angle. "What is a right angle? *How is it defined?"*

A right angle is one that is equal to its supplement. There could be some advantage in restating the conclusion in this sense. Produce the line PO to P' (so that P, O, and P' are collinear, and O and the plane containing O lie between P and P'). Then (see Fig. 7.12*a*) the desired conclusion is

$$\angle POC = \angle P'OC$$

"Why do you think that this form of the conclusion is more advantageous?"

We often prove the equality of angles from congruent triangles. In the present case we could prove the desired conclusion if we knew that

$$\triangle POC \cong \triangle P'OC$$

(see Fig. 7.12*b*). To prove this, however, we should, can, and shall suppose P' so constructed that

$$PO = P'O$$

In fact, what do we need to prove that those two triangles are congruent? We know two pairs of equal sides, $PO = P'O$ by construction, and also

$$OC = OC$$

obviously. To finish the proof it would be enough to know (see Fig. 7.12*c*) that

$$PC = P'C$$

In the foregoing, we have worked starting from the desired conclusion toward the given hypothesis; we have worked *backward.* We went a good distance working backward from the conclusion, although the continuation of the path that should lead us to the hypothesis is still in the clouds. Our work is symbolized by Fig. 7.13, which shows graphically which statements we can conclude from which other statements—as Figs. 7.1 to 7.8 show which quantities we can compute from which other quantities. In Fig. 7.13 each of the foregoing equations (congruences) is represented by its left-hand side:

$$\angle POC = \angle P'OC \qquad \text{by } \angle POC$$
$$\triangle POC \cong \triangle P'OC \qquad \text{by } \triangle POC$$
$$OC = OC \qquad \text{by } OC$$

and so on. It is, in fact, enough to write the left-hand side, because we can derive from it the right-hand side by substituting P' for P, that is, by passing from the space above the plane through A, B, C, and O to the space below it.

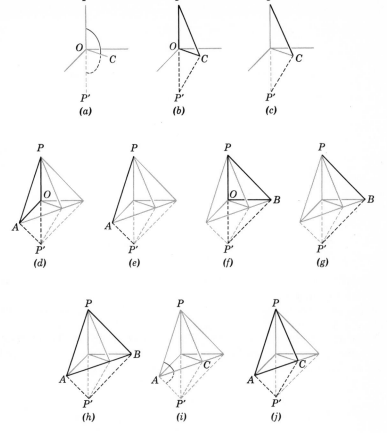

Fig. 7.12. The changing aspect of the geometric figure.

(2) *Restating the problem.* Having worked some time at the conclusion, we should now turn our attention to the hypothesis of the proposition we wish to prove.

"What is the hypothesis?"

We should restate the hypothesis so that it harmonizes with the restated conclusion; we should bring hypothesis and conclusion nearer to each other and not pull them apart. We have to prove (the restated conclusion) that

$$\angle POC = \angle P'OC$$

in supposing (let us restate analogously the hypothesis) that

$$\angle POA = \angle P'OA \qquad \text{and} \qquad \angle POB = \angle P'OB$$

The whole proposition sounds quite good; it appears homogeneous. Yet we have still to add to the hypothesis an essential clause which says that the three differ-

Fig. 7.13. Working backward.

ent lines OA, OB, and OC are in the same plane.   Also this clause should be some-how related to the conclusion.   But how?

In fact, an essential idea is needed here to observe that we are allowed to place the points A, B, and C along a straight line and that it may be advantageous to place them so.   (Any straight line that does not pass through O and is not parallel to any one of the three given lines OA, OB, and OC may be chosen.)   Thus we arrive at a reformulation of the proposition we wish to prove.

○$\angle POC$

○   ○   ○
$\angle POA$ A, B, C collinear $\angle POB$

Fig. 7.14. The gap between the hypothesis and the conclusion.

Hypothesis: The points *A*, *B*, and *C* are collinear, and the line passing through them does not contain the point *O*. Moreover

$$\angle POA = \angle P'OA \qquad \angle POB = \angle P'OB$$

Conclusion: Then

$$\angle POC = \angle P'OC$$

Figure 7.14 symbolizes this statement.

(3) *Working forward.* In working at the hypothesis we consider the same kind of relations as we have considered in working at the conclusion, but we consider them in reverse order.

Since

$$\angle POA = \angle P'OA \qquad \text{by hypothesis}$$
$$PO = P'O \qquad \text{by construction}$$
$$OA = OA \qquad \text{obviously}$$

we conclude that

$$\triangle POA \cong \triangle P'OA$$

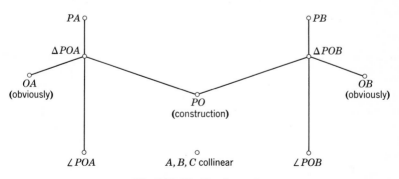

Fig. 7.15. Working forward.

(see Fig. 7.12d) and hence that

$$PA = P'A$$

(see Fig. 7.12e). An exactly parallel reasoning shows that

$$PB = P'B$$

(see Figs. 7.12f and 7.12g).

In the foregoing we have worked *forward;* that, is in the direction from the given hypothesis toward the desired conclusion.

Figure 7.15 exhibits the mental situation resulting from the work just done and from the work recorded by Fig. 7.13. As Fig. 7.15 shows, it remains to prove that $PC = P'C$ from the already proved similar statements $PA = P'A$ and $PB = P'B$ and the heretofore unused hypothesis that A, B, and C are collinear. Comparing the present situation with that represented by Fig. 7.14 we have some reason to be hopeful; the gap that we have to bridge became definitely less wide.

(4) *Working from both sides.* The remaining part of the proof may occur so fast to the problem solver (or to the reader) that the final conclusion may seem instantaneous. Yet let us record the details.

The desired relation

$$PC = P'C$$

(Fig. 7.12c) could be derived from congruent triangles. (This is working backward.) In fact, we can easily derive from the already established two relations

$$PA = P'A \qquad PB = P'B$$

and the self-evident

$$AB = AB$$

the congruence (Fig. 7.12h)

$$\triangle PAB = \triangle P'AB$$

(We have worked forward.) These are, however, not the triangles we need. To derive $PC = P'C$ (which would finish the proof) we should know, for example, that

$$\triangle PAC = \triangle P'AC$$

We could conclude this (now, we are working backward) from the already established

$$PA = P'A$$

and the self-evident

$$AC = AC$$

if we only knew that

$$\angle PAC = \angle P'AC$$

Now, in fact, we know that

$$\angle PAB = \angle P'AB$$

(see Fig. 7.12i) from the already established congruence of the triangles $\triangle PAB$ and $\triangle P'AB$. (Figure 7.16 expresses the mental situation at this instant.) Yet as A, B, and C are collinear, by hypothesis,

$$\angle PAB = \angle PAC \qquad \text{and} \qquad \angle P'AB = \angle P'AC$$

And with this remark we close the last gap (see Fig. 7.17; review the whole Fig. 7.12).

The last step, the transition from Fig. 7.16 to Fig. 7.17 deserves particular attention; only this last step uses the vitally important clause of the hypothesis that A, B, and C are collinear.

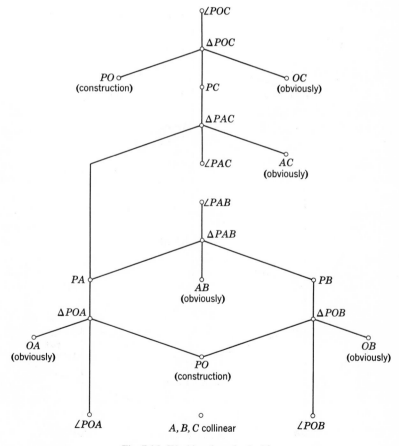

Fig. 7.16. Working from both sides.

7.4. *Elementary diagrams.* In sect. 7.2 to 7.6 we have studied a problem to find, in the foregoing ex. 7.3, a problem to prove. In both cases we have used diagrams consisting of points and connecting lines to illustrate the progress and the structure of the solution. Comparing the two cases, we wish to clarify the meaning of these diagrams.

Let us consider an "elementary diagram," as in Fig. 7.18. This diagram consists of $n + 1$ points of which one, A, is on a higher level than the n others, B, C, D, . . . , and L. The higher placed point A is joined by a line going downward to each of the other n points. Such elementary diagrams are the "bricks" of which the diagrams encountered in Fig. 7.3 to 7.8 and Fig. 7.13 to 7.17 are built up. What does such an elementary diagram express?

When the elementary diagram belongs to a problem to find such as the one considered in Fig. 7.3 to 7.8, the points A, B, C, D, . . . , and L represent *quantities*

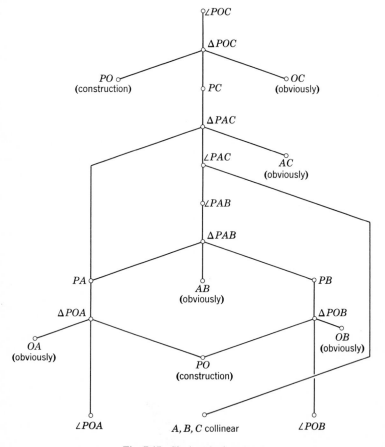

Fig. 7.17. Closing the last gap.

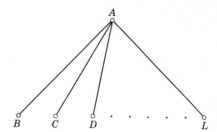

Fig. 7.18. We could have A if we had B and C and D . . . and L.

(lengths, volumes, . . .) and they represent *statements* when the diagram belongs to a problem to prove as in Fig. 7.13 to 7.17. In the former case, Fig. 7.18 expresses that *we can compute the quantity A if we are given the quantities B, C, D, . . . , and L.* In the latter case, Fig. 7.18 expresses that *we can conclude the statement A from the statements B, C, D, . . . , and L.* In other words, the elementary diagram expresses in the first case that the quantity A is a *known function* of the quantities B, C, D, . . . , and L, and in the second case that the statement A is a *consequence* of the statements B, C, D, . . . , and L. We can also say that the elementary diagram Fig. 7.18 answers a question, in one case the question: "From what *data* can we compute A?" and in the other the question: "From what *premisses* can we conclude A?"

And so we can foresee how we could use such diagrams to illustrate the solution of any kind of problem. In a practical problem the points A, B, C, D, and L may represent things we have or we wish to acquire. Figure 7.18 shows that we could acquire A if we had B and C and D . . . and L, or that the means B, C, D, . . . , and L suffice jointly to attain the end A. The diagram answers the question: "What should I get first if I wish to get A?"

7.5. *More problems.* Although we can try to represent diagrammatically the solution of all sorts of problems (see ex. 7.4), the representation can become constrained, unnatural. Seek problems with solutions easily represented and instructively clarified by the diagram.

CHAPTER 8

PLANS AND PROGRAMS

From desire ariseth the thought of some means we have
seen produce the like of that which we aim at; and from
the thought of that, the thought of means to that mean;
and so continually, till we come to some beginning within
our own power.

THOMAS HOBBES: *Leviathan,* Chapter III.

8.1. A pattern of planning

Hobbes' words prefixed to this chapter describe a basically important pattern—the pattern of a procedure of problem solving—with admirable concision and precision. Let us read between the lines, let us try to see the full scope of the procedure, the variety of cases to which it is applicable.

We have a problem. That is we have an aim A which we cannot immediately attain and we are searching for some appropriate action to attain it. This aim A may be practical or theoretical, perhaps mathematical—a mathematical object (number, triangle, . . .) which we wish to find (compute, construct, . . .) or a proposition which we wish to prove. At any rate, we desire to attain our aim A.

"From desire ariseth the thought of some means"—this is well-observed mental behavior. The end suggests the means; a wish is usually quickly followed by the thought of some action that could lead to its fulfillment. I think of some object I would like to have and soon after I remember a shop where I could buy it.

Yet let us return to Hobbes' text: "From desire ariseth the thought of some means" B that could produce the desired A. This thought probably originates from some previous experience: "We have seen B produce the like of A that we aim at." At any rate, we think that we could obtain A if we had B. "And from the thought of B the thought arises of some means, say C, to that mean B"; we could obtain B if we had C. "And so on continually"—we could obtain C if we had D—"till we come to some beginning within our own power"; we could obtain *D* if we had *E*—but we do have *E*! This *E* ends our train of thought; *E* is in our possession, "within our own power," is given, is known.

Our train of thought ran through several ifs—"this if that," we could obtain this if we had that. In fact, we have considered

$$A \text{ if } B, B \text{ if } C, C \text{ if } D, D \text{ if } E$$

and we have stopped at *E* since we have *E* unconditionally, without any further if.

(It is almost unnecessary to mention that the number of ifs, or steps, is unessential; the four steps and five "targets" or "objectives" arising in our example must be considered as representing *n* steps and *n* + 1 objectives.)

The foregoing was *planning*. It should be followed, of course, by the *execution* of the plan. Starting from *E* which is a "beginning within our own power" we should obtain *D*; having obtained *D*, we should proceed to *C*, from *C* to *B*, and finally from *B* to our desired aim *A*.

Observe that planning and execution proceed in opposite directions. In planning, we started from *A* (the aim, the unknown, the conclusion) and we ended by reaching *E* (the things we have, the data, the hypothesis.) Yet in carrying out the plan we worked from *E* to *A* so that *A*, the aim, is the first thing we thought of and the last thing we laid hands on. If we regard motion toward the goal as progress, we must regard the direction in which our planning has moved as *regressive*. Thus the important pattern of problem solving described by Hobbes can be appropriately called *regressive planning,* or *working backwards;* the Greek geometers called it *analysis* which means "backward solution." The complementary work of execution which proceeds from the things in our possession to the aim (from *E* to *A* in our case) is called, in contradistinction, *progressive,* or working forward, or *synthesis* which means "putting together."[1]

The reader should visualize this working backward in planning and working forward in executing the plan on some obvious example. "I could get that desirable object *A* in that shop if I paid for it that sum *B*; I could get that sum *B* if . . . ." I hope that the reader will succeed in devising a neat plan, and will not meet with disappointments in carrying it through.

8.2. A more general pattern

Let us confront the pattern developed in the foregoing section with the example carefully analyzed in chapter 7 and illustrated by Fig. 7.8.

[1] *Cf.* HSI, pp. 141–148, Pappus, and pp. 225–232, Working backwards.

The example unmistakably shows the general trend of the pattern: working backward from the unknown to the data in the planning phase, but working forward from the data to the unknown in executing the plan. Yet the details of the example do not fall under the pattern.

Let us look at the very first step. In the pattern of sect. 8.1, A is reduced to B, the primary aim is reduced to a secondary aim, attaining A depends on attaining B. In the example of Fig. 7.8, however, the computation of the unknown (the volume of the frustum) is reduced to the computation of *two* new unknowns (two volumes); there is not only one secondary aim, but there are two such aims.

If, however, we reconsider our example represented by Fig. 7.8 and various remarks made in chapter 7 on this graphical representation (see especially ex. 7.4), it should not be difficult for us to conceive a generalization of the pattern of sect. 8.1 that encompasses the case of Fig. 7.8 and along with it an unlimited variety of worthwhile cases.

Our aim is A. We cannot attain A immediately, but we notice that we could attain it if we had several things, B', B'', B''', ... Well, we do not have these things, but we start thinking how we could get them; that is, we set up B', B'', B''', . . . as our secondary aims. We may perceive after some reflection that we could attain all our secondary aims B', B'', B''', . . . if we had several other things C', C'', C''', ... In fact, we do not have these things $(C', C'', C''', . . .)$, but we can try to get them; we set them up as our tertiary aims, and so on. We are spinning the web of our plan. We may be obliged to say several times "We could have this, if we had that and that and that" till we finally reach solid ground, things we really have. The web of our plan consists of accessory aims, all subordinate to our primary aim A, and of their interconnections. There may be many subordinate aims, and the details of the web may be too complex for words, but they can be appropriately represented by the points and lines of a diagram of the kind we developed in chapter 7. (For instance, in sect. 2.5(3) our primary aim was D, our secondary aims were a, b, and c, and our tertiary aims p, q, and r. See also ex. 7.2.)

I think that the foregoing indicates clearly enough a general pattern which contains the pattern described in sect. 8.1 as a particular case, and which we shall call the *pattern of working backwards*. It is a pattern of planning. It is a pattern of planning; the planning starts from the aim (the thing we want, the unknown, the conclusion) and works backward toward the things "within our power" (the things we have, the data, the hypothesis.) It is part of the plan that when we shall have reached those things "within our power" we shall use them as a "beginning," and by retracing our steps we shall work forward toward the aim. *Cf.* ex. 9.2(3).

8.3. A program

Are the two numbers $\sqrt{3} + \sqrt{11}$ *and* $\sqrt{5} + \sqrt{8}$ *equal? If they are not, which one is greater?* (It is understood that all arising square roots are taken with the positive sign.)

With a little experience in algebraic manipulation, we may readily conceive a plan to handle this question; we may even conceive a plan so clear and determinate that it deserves a special term and may be called a *program.*

The two proposed numbers are either equal, or the first is greater, or the second is greater. There are three eligible relations between the two numbers, expressed by the signs $=$, $>$, and $<$, but just one of these three relations is actually valid—we do not know for the moment which one, although we hope that we shall know it soon. Let us denote the one among the three relations that is actually valid by ? and write

$$\sqrt{3} + \sqrt{11} \quad ? \quad \sqrt{5} + \sqrt{8}$$

Whichever of the three relations may be actually valid, we can perform certain algebraic operations equally applicable to all three. To begin with, we can square both sides and then the same relation will hold between the squares:

$$3 + 2\sqrt{33} + 11 \quad ? \quad 5 + 2\sqrt{40} + 8$$

By this operation we have diminished the number of square roots; at the outset we had four, and now we have only two. By subsequent operations we shall get rid of the remaining square roots by and by, and then we shall see which one of the three possible relations is represented by the sign ?.

The reader need not foresee the operations planned in every detail, but he should realize that they can be carried out without hesitation and are bound to lead to the desired decision. Then he may agree that the situation deserves a special term and that such a definite plan should be called a program, (see sect. 8.5).

With these remarks we have, in fact, attained the aim of the present section and there is no need to carry out the programmed steps. Still, let us carry them out:

$$1 + 2\sqrt{33} \quad ? \quad 2\sqrt{40}$$
$$1 + 4\sqrt{33} + 132 \quad ? \quad 160$$
$$4\sqrt{33} \quad ? \quad 27$$
$$528 \quad ? \quad 729$$

Yet there is no question here, we know which side is larger, and retracing our steps we find that

$$\sqrt{3} + \sqrt{11} < \sqrt{5} + \sqrt{8}$$

8.4. A choice between several plans

On each side of a given (arbitrary) *triangle describe an equilateral triangle exterior to the given triangle, and join the centers of the three equilateral triangles.* Show that the triangle so obtained is EQUILATERAL.

In Fig. 8.1, $\triangle ABC$ is the given triangle, and A', B', and C' are the centers of the equilateral triangles described on BC, CA, and AB, respectively. We are required to show that $\triangle A'B'C'$ is equilateral, but it seems odd, almost unbelievable, that this triangle is always equilateral, that the final shape produced by the construction is independent of the arbitrary initial shape. We suspect that the proof cannot be easy.

At any rate we do not like the points A', B', and C' appearing so isolated from the rest of Fig. 8.1. This defect, however, is not serious. As it is readily seen, $\triangle BA'C$ is isosceles, $A'B = A'C$, and $\angle BA'C = 120°$. We introduce this triangle and two analogous triangles into the figure and thus obtain the "more coherent" Fig. 8.2.

Yet we still do not know how to approach our goal. *How can you prove such a conclusion?* In Euclid's manner? By analytic geometry? By trigonometry?

(1) How can we prove in Euclid's manner that $A'B' = A'C'$? By showing that $A'B'$ and $A'C'$ are corresponding sides in congruent triangles. Yet there are no usable triangles in the figure, and we do not see how we could introduce usable triangles. This is discouraging—let us try another approach.

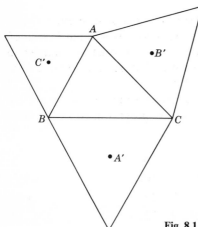

Fig. 8.1. Three isolated points.

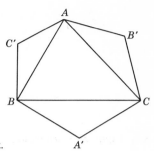

Fig. 8.2. More coherent.

(2) How can we prove by analytic geometry that $A'B' = A'C'$? We should regard the coordinates of the points A, B, and C as given quantities and the coordinates of the points A', B', and C' as unknowns. Having expressed these unknowns in terms of the data, we can express also the distances in question in terms of the data and examine whether they are equal or not. It is a pretty clear plan, but we should handle six unknowns and six data—no, it is not too inviting—let us try the third approach.

(3) How can we prove by trigonometry that $A'B' = A'C'$? We should regard the sides a, b, and c of $\triangle ABC$ as given quantities, and the three distances

$$B'C' = x, \qquad C'A' = y, \qquad A'B' = z$$

as unknowns. Having computed the unknowns, we should examine whether actually $x = y = z$. This looks better than (2); we have only three unknowns and three data.

(4) In fact, we need not compute three unknowns, two are enough: if $y = z$, *any* two sides are equal—and that is enough.

(5) In fact, we need not even compute two unknowns, one is enough if we proceed a little more subtly; it is enough to express just x in terms of a, b, and c provided that we manage to obtain an expression *symmetric* in a, b, and c. (An expression is symmetric in a, b, and c if it remains unchanged when we interchange a, b, and c. If such an expression is valid for x it must be valid just as well for y and z.)

This plan, although it depends on the ingenuity of the problem solver, on the coming of some little novel idea, appears rather attractive—the reader should try to carry it out (see Fig. 8.3, ex. 8.3).

(6) Has our story a moral? I think, it has.

If you see several plans, none of them too sure, if there are several roads diverging from the point where you are, explore a bit of each road before you venture too far along any one—any one could lead you to a dead end.

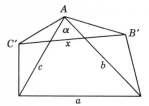

Fig. 8.3. Concentrate on one side.

8.5. Plans and programs

We may regard our plan as a road along which we intend to travel. There are plans and plans, however. We would like to have a plan of some action that leads us right to the goal, but, unfortunately we do not always succeed in devising such a complete plan, and also less far-reaching actions need some planning. We may see just a little bit of the road, we may see a long stretch of it, or we may see the full extent of our road right to the goal. Then, we may see our road hazily or clearly. Again, we may expect, and prepare for, various eventualities along the portion of the road that we do not see well or do not see at all. A desirable eventuality, the hope for which never leaves us entirely, is to have a bright idea which will immediately clear up everything.

Quite often, we do not have yet a complete plan; there are gaps in the plan, some vital ideas are still missing. Yet we go ahead, we start executing our plan in relying on some bright, or at least novel, idea that will come to us and fill the gaps.

How much we rely on such an idea may be regarded as the most important distinction between plans and plans. If we do not depend at all on novel ideas, but are confident that the steps already pondered and foreseen by the plan are sufficient to reach our goal, we have a plan clear and determined enough to be called a *program*. We may spend much of our time in working at various imperfect plans till we succeed in developing one of them into a program.

Compare sect. 8.3 with sect. 8.4, for instance.

8.6. Patterns and plans

Under appropriate circumstances, each pattern we have studied in the foregoing suggests a plan—but does not immediately yield a definite plan, a program.

For instance, we have a problem of geometric construction. We may try to solve it following the pattern of two loci. This is a plan, indeed, but it takes additional ideas to find a suitable point to the construction of which the problem can be reduced, and to split the condition suitably so that we obtain two loci for that point.

Or, we decide to solve a geometric problem by reducing it to equations, following the Cartesian pattern. This is a plan, indeed, but it takes additional ideas to set up as many equations as unknowns and further ideas to solve the system of equations.

Working backward is a very general and useful pattern of planning but obviously we need some ideas from the subject matter to work across the gap from the unknown to the data. When the regressive planning has succeeded and the logical network spanning the gap has been perfected, the situation is very different. Then we have a program for working foreward from the data to the unknown.

Examples and Comments on Chapter 8

8.1. *Backward or forward? Regressive or progressive? Analysis or synthesis?* In our terminology (see sect. 8.2) the term "working backward" stands for a certain strategy of problem solving, a typical way of planning the solution. Is it the only possible strategy? Is it the best strategy?

(1) Take "our example," the example we made a graphic study of in chapter 7. The plan of the solution at which we have finally arrived is represented by Fig. 7.7; it is a web of points and lines, of intermediate unknowns and interconnections, stretching across the originally open space between the unknown and the data. We started spinning this web from the unknown, working toward the data; the successive stages of the work are displayed in Fig. 7.8. We called this direction of work "regressive" or "backward" (in Fig. 7.8 it is downward).

The final plan, the full system of interconnections (see Fig. 7.7—the web may be more complex in another case) does not show in which direction it has been built up. Another problem solver could have built it up starting from the data and working in the direction of the arrows in Fig. 7.7 (which we have followed in executing the plan). Discovering the plan in this direction would be *progressive* work, *working forward.*

Still another problem solver (having, perhaps, a more complicated problem) could devise such a plan without working all the time in the same direction. Starting from either end, he could work sometimes from the unknown toward the data, sometimes from the data toward the unknown; he may work alternately from both ends; he may even establish some promising connection in the middle between things which are not yet connected with either end. Thus, devising a plan in working backward is by no means the only possibility. For a concrete case, see ex. 7.3.

(2) In our example summarized by Fig. 7.8 we have devised the plan of the solution working backward. Let us compare our work with the work of a problem solver who happened somehow to devise the same plan working foreward.

In fact, we have started from the unknown, we have asked ourselves questions emphasizing the unknown: *What do you want? What is the unknown? How can you get this kind of thing? How can you find this kind of unknown? From what data can you derive this kind of unknown?* And thus we have found two "data," the volumes A and B, from which the unknown can be derived, in terms of which the unknown F can be expressed: $F = B - A$. This stage of our work is indicated by Fig. 8.4 (which is a part of Fig. 7.3).

Yet the other problem solver started differently: from the data. He asked himself questions emphasizing the data: *What have you? What are the data? What are such things good for? How can you use such data? What can you derive from such data?* And somehow he observed that he can derive from the data the length (altitude) x, express x in terms of a, h, and b (from a proportion as we have observed too later in the game, see Fig. 7.6). This stage of his work is indicated by Fig. 8.5.

Let us return to our work, to the stage represented by Fig. 8.4. Having expressed the unknown F in terms of A and B, we are left with two new unknowns, A and B, two new (auxiliary) problems:

> Compute A in terms of the data a, h, and b.
> Compute B in terms of the data a, h, and b.

These are clear-cut mathematical problems, of the same kind as our original problem. Working regressively, we ask again: *How can you find such unknowns? From what data can you derive such unknowns?*

Let us now return to the other problem solver; he has reached the stage represented by Fig. 8.5. Having expressed x in terms of the data a, h, and b, he can regard x as given, and so he has more data which may give him a better chance to find the original unknown. Progressing in the same way, however, he has no clear-cut auxiliary problem, but he has to ask himself the less definite question: How can I use x? *What are such things good for? What can I derive from such data?*

Yet the most striking difference between us and the other problem solver, between the two situations represented by Fig. 8.4 and 8.5, is in the outlook. What happens if we succeed in solving our auxiliary problems? What happens if he succeeds in answering his question?

If we succeed in expressing our auxiliary unknowns in terms of the data (A and B in terms of a, h, and b), we can express so also our original unknown

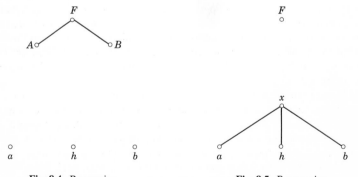

Fig. 8.4. Regressive.   **Fig. 8.5.** Progressive.

$(F = B - A)$ and our problem is solved. If the progressive worker succeeds in expressing some quantity, say y, in terms of the data, he still faces an indefinite question: How can he use y? Except for one case, of course: if he is lucky, he can express F itself, and then he also has solved his problem.

(3) Both ways of planning, the regressive and the progressive, can succeed, and both can fail. Working backward, we may arrive at an auxiliary problem which we cannot solve. Working foreward, we may derive more and more quantities from the data, but these quantities may serve no useful purpose; we may not be able to derive from them our original unknown.

Both ways of planning require a combination of various activities. Yet, when working backward we may expect to spend most of our time in doing clear cut problems. When working foreward, we may expect to spend much of our time in hesitating between problems we might do, or in doing problems which are no help.

On the whole, planning regressively, working backward, "analysis" (in the terminology of the Greek geometers) is preferable. There can be no hard and fast rule, but the wise thing may be to look first at the unknown (the conclusion, the thing you want) then at the data (the condition, the hypothesis, the things you have). Start working backward from the unknown unless you have some special reason not to do so—of course, if a bright idea urges you to start from the data, go ahead!

(4) Let us make just a few more short remarks, although there are still many things to say.[2]

In some cases there is little choice. In many a practical problem the thing we want to find (to construct, to acquire, . . .) is quite definite, but the things we could possibly use for attaining our aim we scarcely know, and they are impossible to survey, there are so many of them. We can hardly have a good reason to begin with any one item in that unmanageable heap of data, and so we are almost obliged to plan regressively.

After arriving at our plan regressively, we execute it progressively (recall sect. 7.5); but it is execution and not planning for we have conceived all the ideas before, and now we just carry them out. This may even raise the suspicion that those people who start planning progressively may have conceived some idea before—conceived it, I mean, implicitly, perhaps subconsciously.

A lady student explained it thus: synthesis by itself (without foregoing analysis) would be difficult—somewhat like trying to make a cake with the ingredients in front of you and no recipe to follow.

And, of course, you should not be pedantic or fussy. Having started working backward, from the unknown, you may see a good opportunity to take a step forward from the data; then take that step by all means.

8.2. A wise man begins in the end. A friend of mine, a good mathematician and a good philosopher, once told me that when he tries to find a proof for a theorem he often begins by writing down Q.E.D. ("quod erat demonstrandum," what was required to prove), and this act of writing down the traditional phrase which comes at the very end of a proof puts him in the right mood.

There is a proverb: "A wise man begins in the end, a fool ends in the beginning."

---

[2] Cf. HSI, pp. 141–148, Pappus, and pp. 225–232, Working backwards.

8.3. Carry out the plan indicated in sect. 8.4(5).

8.4. *A choice between three plans.* Let a denote the radius of the base and h the altitude of a right circular cylinder. A plane intersects the base along a diameter, is tangent to (has just one point in common with) the perimeter of the top, and divides the volume of the cylinder into two unequal portions. Compute the smaller portion (the volume of a "hoof") which is between the base and the intersecting plan.

This problem and its first solution are due to Archimedes.[3]

We use solid analytic geometry. The axis of the cylinder is the z-axis, and its base lies in the x, y-plane of a rectangular coordinate system. The plane dividing the volume of the cylinder intersects the x, y-plane in the y-axis. Therefore, the equation of the perimeter of the base is

$$x^2 + y^2 = a^2$$

and the equation of the dividing plane is

$$\frac{z}{h} = \frac{x}{a}$$

We may use either integral calculus or Cavalieri's principle to compute the desired volume. With both methods we have to consider a family of parallel cross sections of the "hoof." There are three obvious plans: we may choose cross sections perpendicular to the x-axis, perpendicular to the y-axis, or perpendicular to the z-axis.

Which plan do you prefer? Carry it out.

8.5. *A choice between two plans.*

(1) In doing a crossword puzzle we hesitate between two words. One has 4 letters, of which 1 is known and 3 are unknown, the other has 8 letters of which 3 are known and 5 are unknown. Which one of these two words should we try to find first? *On the basis of the numerical data* offered, is it possible to make a rational choice between the two?

It is hardly possible, I think, but there is a challenge.

†(2) The question raised must be restated in a more general and (as far as possible) more precise form.

A word consists of $k + l$ letters of which k are known and l are unknown. We undertake to find the word, and we are considering the *degree of difficulty* of this undertaking.

Let us first suppose that we know those k letters *completely:* the nature and position of each of them are specified (as in the example

$$\text{IN} \_ \_ \text{R} \_ \_ \_$$

where $k = 3, l = 5$). In this case we may define the degree of difficulty of finding the word as the number N of those English words in present-day usage that consist of $k + l$ letters of which k letters are of the specified nature and are in the specified position. (Of course, any well-defined increasing function of N, such as $\log N$, could be just as well taken for the degree of difficulty.)

---

[3] See Proposition 11 of his *Method,* pp. 36–38 of the Supplement to his *Works* edited by T. L. Heath (Dover.).

Theoretically, this definition may appear reasonable; the greater the number N of the admissible words, the greater is the difficulty to choose one of them. Practically, there are several awkward points. How should we decide whether an English word is or is not "in present-day usage"? Is the definition satisfactory from the standpoint of the crossword fan? At any rate, the practical determination of the number N appears prohibitively tiresome and unprofitable.

†(3) Thus we are led to a different and higher aim; we wish to determine the degree of difficulty *insofar as* it depends on k and l alone—the difficulty "other things being equal"—perhaps the "average difficulty." We wish to lump together all the cases with the same k and l, and take only the numerical data k and l into account. If we could attain this higher aim, the degree of difficulty would be a numerical function $f(k, l)$ of k and l. It seems obvious that $f(k, l)$ should be a decreasing function of k and an increasing function of l. Yet, for instance, we do not know yet which one is greater, $f(1, 3)$ or $f(3, 5)$.

†(4) If the letters in English words were *independent* of each other, the number N of English words with k given and l freely eligible letters could be simply expressed as follows:

$$N = 26^l$$

Number N is used in the meaning explained under (2). Hence the degree of difficulty could be defined, for instance as

$$f(k, l) = \frac{\log N}{\log 26} = l$$

This choice of $f(k, l)$ is consistent in itself, but it sidesteps the crucial question: How much do the k known letters restrict the choice of the l still eligible (but, in fact, not quite freely eligible) letters?

It is very doubtful that a somewhat realistic formula for $f(k, l)$ can be devised. At any rate, we expect that such a formula would differ from the one just proposed at least in two respects: it should be a strictly decreasing function of k, and it should be adaptable, if not to all, at least to a few languages.

†(5) Here is a crude and completely speculative trial proposal:

$$f(k, l) = \frac{\log [26 - \alpha k][26 - \alpha(k + 1)] \cdots [26 - \alpha(k + l - 1)]}{\log 26}$$

Suitable choice of the positive parameter α should adapt the formula to a language written with the 26 Roman letters. The formula should be used only for words of length

$$k + l < \frac{25}{\alpha} + 1$$

(6) The foregoing considerations may shed some light on the scope of heuristic and on the precision that it could possibly attain, and herein lies their justification.

8.6. *A plan, indeed.* "I will just sit here, look at the figure, and wait for a good idea." This, indeed, is a plan. Perhaps a little too primitive. Perhaps a little too optimistic: you seem to rely just a little too much on your ability to produce good ideas. Still, such a plan may work, occasionally.

8.7. Looking back at problems you have solved in the past, reconsider some problems you have solved, or *could* have solved, in working backward.

8.8. *Do not commit yourself.* We consider a half-concrete example. We are required to prove a proposition of elementary geometry the conclusion of which states, "... then the angles ∠*ABC* and ∠*EFG* are equal." We have to derive this conclusion from a certain hypothesis the details of which, however, are irrelevant to the purpose of this comment, and so we ignore those details.

At a certain (probably early) stage of the problem solving process we concentrate on the conclusion: *What is the conclusion?*

We have to prove that

$$\angle ABC = \angle EFG$$

How can you prove such a conclusion? From what hypothesis can you derive such a conclusion?

We succeed in remembering several pertinent facts learned in the past; several possibilities of deriving such a conclusion as we have to prove. Two angles are equal,

(1) if they are corresponding angles in congruent triangles; or

(2) if they are corresponding angles in similar triangles; or

(3) if they are corresponding angles formed by two parallels with a transversal; or

(4) if they have the same third angle as complement; or

(5) if they are inscribed in the same circle and intercept the same arc.

We have here five different known theorems, each of which could be applicable to our case, five different hypotheses from each of which we could derive the de-

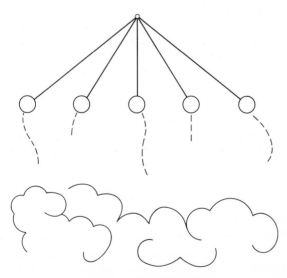

Fig. 8.6. A doubtful choice.

sired conclusion. We may start working with any one. For instance, we could try (1); we could introduce two likely triangles, say $\triangle ABC$ and $\triangle EFG$, and then try to prove that they are congruent. If we succeeded, the desired conclusion would immediately follow! And how could we prove that $\triangle ABC \cong \triangle EFG$? With this question we embark upon regressive planning. Yet we could begin such regressive planning just as well with any one of the other theorems quoted— has any one of them some chance of success? Which one has the best chances of success? If we cannot answer these questions, if the response they evoke is just some dim uncertain feeling, the choice we are facing is doubtful indeed. We are at the crossroads. We have to choose between several paths; the beginning of each is clear enough, but the continuation is uncertain and the end is hidden by clouds. Figure 8.6 attempts to express the situation graphically.

The aim of this example was just to make the reader realize the perplexing situation, the uncertainty of the choice between several plans. The advice I would give in such a situation is: *Do not commit yourself too early, do not commit yourself to one course more than necessary.* Do one thing, but do not forget the others.

The good problem solver plans like a good general; he realizes that the attack planned may fail and does not neglect the line of retreat. A good plan must have some built-in *flexibility*, some adaptability to unforseeable difficulties.[4]

[4] See MPR, Vol. 2, pp. 148–152.

CHAPTER 9

PROBLEMS WITHIN PROBLEMS

When, in either construction or demonstration, we assume anything which has not been proved but requires argument, then we regard what has been assumed as doubtful in itself and worthy of investigation, and call it a lemma.

> PROCLUS: *Commentary on Euclid,* On Proposition 1 of Book I.

When a problem arises, we should be able to see soon whether it will be profitable to examine some other problems first, and which others, and in which order.

> DESCARTES: *Œuvres,* vol. X, p. 381; Rules for the Direction of the Mind, Rule VI.

What is the best you can do for this problem? Leave it alone and invent another problem.

> THE TRADITIONAL MATHEMATICS PROFESSOR.

9.1. Auxiliary problems: means to an end

Some observations of Wolfgang Köhler on anthropoid apes have great interest for us. Let us describe, schematically, one of his experiments.[1]

Inside the cage there is a chimpanzee and he is hungry. Outside the cage there lies a banana on the ground. The chimpanzee can pass his arms through the bars of his cage, but the banana is beyond his reach. The chimpanzee has tried hard to reach the banana, but without

---

[1] Wolfgang Köhler, *The mentality of apes* (Harcourt, Brace & Co.) pp. 32–34.

36

success, and now he just sits there. Also on the ground outside his cage, and within his reach, there lies a stick, but he seems to pay no attention to it. Suddenly he stirs, grabs the stick, clumsily pushes the banana with it until he can reach the banana, and then he grabs it and eats it. This chimpanzee has solved two problems:

A. To grab the banana.
B. To grab the stick.

Problem A arose first. Originally, the chimpanzee did not show the least interest in the stick, which he cannot eat; yet he solved B first. The solution of problem B paved the way to the solution of his original problem A. The chimpanzee had a direct interest in A and only an indirect interest in B; A was end, B only means to him; A was his main, or original, problem, B just an *auxiliary problem* ("helping" problem, subproblem).

Let us outline generally the meaning of this important term: *An auxiliary problem is a problem on which we spend attention or work not for its own sake, but because we hope that such attention or work may help us to solve another problem, our original problem.* An auxiliary problem is means to an end, it should yield access to the goal; the original problem is the end and the goal.[2]

Gaining access to the solution of an apparently inaccessible problem by devising and solving first an appropriate auxiliary problem is the most characteristic kind of intelligent action. We can hardly refuse to regard the performance of the chimp as an act of intelligence.

We are going to classify auxiliary problems, starting from mathematical examples.

9.2. Equivalent problems: bilateral reduction

We begin with an example. Our task is to solve the following system of three equations with three unknowns:

$$(A) \begin{cases} x - y \quad\;\; = -4 \\ x + y + z = 5 \\ x + y - z = 31 \end{cases}$$

From system (A) we derive another system (B):

(1) we leave the first equation of (A) unchanged;
(2) we add the second and third equations of (A);
(3) we subtract from the second equation of (A) its third equation

[2] HSI, pp. 50–51; Auxiliary problem.

and so we obtain the three equations of a new system:

$$(B) \begin{cases} x - y = -4 \\ 2(x + y) = 36 \\ 2z = -26 \end{cases}$$

Our derivation of (B) shows that such numbers x, y, z as satisfy (A) must, of necessity, satisfy (B). The converse is also true: numbers x, y, z satisfying (B) must satisfy (A). This seems plausible, but we can also prove it in various ways, for instance as follows. By dividing the last two equations of (B) by 2, we obtain

$$(C) \begin{cases} x - y = -4 \\ x + y = 18 \\ z = -13 \end{cases}$$

and from (C) we can return to (A) by leaving the first equation of (C) unchanged and first adding, then subtracting, the last two. In brief, if three numbers x, y, and z satisfy one of the two systems (A) and (B), they must of necessity satisfy the other.

The systems (A) and (B) are not identical; they do not consist of the same equations. Therefore, we can not properly say that the two corresponding problems, the one to solve (A), the other to solve (B), are identical. Yet we can properly say that these two problems are *equivalent*. Here is the general definition of this use of the term: *Two problems are equivalent for us if we know that the solution of each involves the solution of the other*.[3]

The transition from a problem to an equivalent problem is called a *bilateral* (or reversible, or convertible, or equivalent) *reduction*. For instance, the transition from our original problem which was to solve (A) to the problem to solve (B) is a bilateral reduction. It is a useful reduction too; the system (B) is closer to the solution than the system (A). In fact, (B) is closer to (C) than (A), and (C) is almost at the end of our task; there is already the value of z, and little remains to do to get also the values of x and y.

9.3. Chains of equivalent problems

Let us return to system (C) in the sect. 9.2; we derive from it, by addition and subtraction, the system

$$(D) \begin{cases} 2x = 14 \\ 2y = 22 \\ z = -13 \end{cases}$$

---

[3] HSI, p. 53; Auxiliary problem 6.

and hence

$$(E) \begin{cases} x = 7 \\ y = 11 \\ z = -13 \end{cases}$$

We have here a sequence of five systems (each one is a system of three equations)

$$(A), \quad (B), \quad (C), \quad (D), \quad (E)$$

There is a problem associated with each: to find the values of x, y, and z satisfying the system. [The "problem" is fully solved, and so the term "problem" is not used in its proper, but in an extended, meaning in the case of the system (E).] Each of these problems is equivalent to the foregoing (and also to the following) problem as each link of a chain is joined to the next link; we have here a *chain of equivalent problems*.

In our chain, (A) is the beginning and (E) is the end; (A) is the originally proposed system of equations and (E) exhibits the solution. We have here an ideally perfect way to arrive at the solution. Starting from the proposed problem, we devise a sequence of problems; each problem is equivalent to, and nearer to the solution than, the foregoing problem; proceeding thus from problem to problem, we attain, with a last step, the solution itself.

Yet, even in mathematics, in the quest of the unknown and in the struggle for the proof, we often have to settle for something less than perfection. And so we turn to survey further kinds of auxiliary problems.

9.4. More, or less, ambitious auxiliary problems: unilateral reduction.

We begin with the consideration of a schematically stated problem:

A. Find the volume of a pyramid, being given . . . .

We suppose here that the data are sufficient to determine the pyramid, but the base and the altitude are not among the data, neither of these two quantities is given. This much is important, but what the data otherwise are is unimportant for our present discussion and so we suppress their list.[4]

We know that the volume of a pyramid can be computed if its base and altitude are given—but, as we have just said, neither of these two quantities is given. As they are not given, we shall try to compute them, and so we turn to another problem:

---

[4] For a concrete problem of the form A see ex. 4.17.

B. Find the base and the altitude of a pyramid, being given . . . .

Problem A has one unknown, problem B has two unknowns, and both problems have the same (unlisted) data. There is a one-sided asymmetric relation between these two problems. If we succeed in solving B, we have the base and the altitude of the pyramid, hence we can compute its volume, and so we can solve A. If, however, we succeed in solving A, it is by no means certain that we can also solve B: although the result of A yields a simple relation between the two unknowns of B, there may remain some serious difficulty to find either of them. Thus, we achieve less in solving A than in solving B. We may call A the *less ambitious* and B the *more ambitious* of the two problems.[5]

Let us repeat the foregoing in general terms. There are two problems, A and B, both unsolved, and our state of knowledge is as follows: *we do know how we could derive from the solution of* B *the solution of* A, *but we do not know how we should derive from the solution of* A *the solution of* B. Under such circumstances we say that A is less ambitious than B and that (which means the same) B is more ambitious than A.

The transition from an original problem to an auxiliary problem which is more ambitious, or less ambitious, than the original problem (in any case, not equivalent to the original problem) is called a *unilateral* (or irreversible) *reduction*. In our example, the original problem A is less ambitious than the auxiliary problem B, and so the reduction of A to B is unilateral. The experienced reader may recall several examples similar to the one given here in which such a unilateral reduction was profitable.

Unilateral reduction of the opposite kind, where the auxiliary problem is less ambitious than the original problem, is also often profitable. Here is a schematic example:

A. Compute the unknowns $x_1, x_2, \ldots x_{n-1}$, and x_n, being given . . . .
B. Compute the unknown x_1, being given . . . .

We suppose that the condition and the data determine the unknowns and that they are the same in both problems, in A and in B, but what they are is unimportant here and so we suppress them. Trivially, the solution of A involves the solution of B, but, in general, the solution of B cannot involve the solution of A: according to our definition, A is more ambitious than B. Yet, very often when we are required to solve A, we may introduce with advantage B as auxiliary problem; we did so many times in chapter 3 when we solved A by recursion and, by choosing x_1 as our initial unknown, we began our work with the auxiliary problem B as a stepping stone to A.

[5] HSI, p. 56; Auxiliary problem 8.

9.5. Remoter auxiliary problems

We begin with an example. Let us consider the following problem:

A. Given the length of an edge of a regular tetrahedron, find the radius of the sphere circumscribed about the tetrahedron.

If we do not see some other access to problem A we may try to approach it by considering the following problem:

B. Given the length of a side of an equilateral triangle, find the radius of the circle circumscribed about the triangle.

The transition from A to B is neither a bilateral nor a unilateral reduction in the sense of sections 2 and 4. In fact, we can scarcely see *a priori* how the solution of B would involve that of A or the solution of A that of B—problems A and B do not appear equivalent, and neither appears more ambitious than the other in the sense of our definitions.

Yet problems A and B are not unrelated. Problem B is analogous to the problem A; we have here a little example of that great analogy between plane geometry and solid geometry. Moreover, to most of us the problem B will appear easier than the problem A; we may have even seen already, and could recall with little trouble, the solution of B. In this situation, it is natural to ask: Is it worthwhile to consider problem B? Is there a chance that the consideration of B will facilitate the solution of A?

It may happen that the consideration of B does not contribute appreciably to the solution of problem A; this may even happen if we clearly see the analogy between A and B and if we possess a complete solution of B. Yet it may also happen that B will help, although it is apparently sterile. The comparison of A with the analogous B may render the proposed problem A more interesting, and in such a case B *is* useful. Yet the contribution of B to the solution of A may be even more distinct; there is a chance that the analogy between A and B will suggest some useful remark. For instance, in the "plane" problem B, the desired radius is a simple fraction ($\frac{2}{3}$) of the altitude of the equilateral triangle. This may suggest the question: How about the analogous "solid" problem A? Is that desired radius some simple fraction of the altitude of the regular tetrahedron? This question, or a similar question, can introduce some usable element and pave the way to the solution of A. Or, in solving A, we may need the altitude of one of the faces of a tetrahedron; then, if we know the ratio of the altitude to the radius of the circumcircle for an equilateral triangle (which we have just mentioned) the answer to problem B may contribute a link to the chain that we have to forge to obtain the answer to A.

In general, we may expect that the consideration of a problem B will contribute in some way or other to the solution of the proposed problem A, even if B is neither equivalent to A, nor more ambitious, nor less ambitious, than A. Such a problem B is called a *remoter* auxiliary problem to A.

9.6. Material help, methodological help, stimulating influence, guidance, practice

An auxiliary problem may help to solve the original problem in an inexhaustible variety of ways.

An equivalent auxiliary problem yields, if solved, the *whole solution* of the original problem, and an auxiliary problem more ambitious than the original problem does the same. (The difference between these two kinds of auxiliary problems shows up when we are unable to solve them. If an equivalent auxiliary problem is definitely beyond our reach, so is our original problem; if, however, a more ambitious auxiliary problem turns out inaccessible, the prospects for our original problem need not be so dark.)

Other sorts of auxiliary problems, even if solved, do not guarantee the whole solution of the original problem, but they may offer *material help*. A part of the solution of the auxiliary problem (or even all of it) may become a part of the solution of the original problem to which it may contribute a conclusion, a construction, or a fact on which such a conclusion or construction is based, and so on.

Yet even when no such material help is forthcoming, the auxiliary problem may give *methodological help;* it may suggest the method of solution, an outline of the solution, or the direction in which we should start working, and so on. An auxiliary problem analogous to, but easier than, the original problem is in a good position to offer such methodological help.

We might not be able to point out any part or feature in the final solution of the original problem that was taken over from, or suggested by, a certain auxiliary problem. Still, it is quite possible that the *stimulating influence* of that auxiliary problem contributed a good deal to the discovery of the solution of the original problem. Perhaps that auxiliary problem has rendered, by analogy or contrast, the original problem more understandable or more interesting; or perhaps it has stirred our memory—started a train of thought from which, eventually, some essential relevant fact emerged.

Auxiliary problems may help in still another rather subtle manner. Working on a problem involves decisions. We could continue our work in two directions; there are two paths open for us, one to the right and one to the left. Which one should we choose? Which one is more likely to lead us to the solution? It is important to assess our prospects reasonably and auxiliary problems may give us welcome guidance in this respect. Our attention and work spent on, and our experience gained with, an auxiliary problem have a good chance to influence our judgment in the right direction.

Sometimes we may take on an auxiliary problem just for *practice*. It may happen that our original problem involves concepts with which we are not used to dealing. In such a situation, it may be advisable to try some easier problem involving the same concepts that would thus become a (rather remote) auxiliary problem to our original problem.

Although there are so many different things to gain, it quite often happens that we gain little and lose a lot of time and trouble in working on an auxiliary problem. Therefore, before we get too deeply involved in such a problem, we should try to weigh the prospects and estimate the chances.

Examples and Comments on Chapter 9

9.1. *Reliable sources of auxiliary problems?* An auxiliary problem may "spontaneously grow out" of the proposed problem. Yet it may also happen that when we would like to have an attractive auxiliary problem none comes to mind. In such a case we may wish we had a list of reliable sources from which useful auxiliary problems could be drawn. There are, in fact, various usual modes of forming auxiliary problems, and we shall consider the most obvious ones in what follows; they will lead you to some auxiliary problems in most cases—but it cannot be guaranteed that they will lead you to useful auxiliary problems.

Auxiliary problems may arise at any stage of the problem solving process. Let us assume, however, that we are a little beyond the very first phase. We have already considered and well understood the principal parts of our problem—the unknown, the data and the condition, or the hypothesis and the conclusion—and also the most obvious subdivisions (clauses, etc.) of these principal parts. Yet we see no promising plan, and we wish we had some more accessible or more attractive goal. It is good to know that searching examination of the principal parts of our problem may present us with such a goal, with a usable auxiliary problem. We shall survey the most notable cases in what follows.

9.2. *Respice finem.* The desire to attain the aim is a productive desire, it produces thoughts of actions which possibly could attain the desired aim. The end suggests the means. Therefore, look at the end, do not lose sight of your aim; it guides your thoughts.

Respice finem means "Look at the end" and was a current phrase when Latin was more used.[6] Hobbes expands it: "Look often upon what you would have, as the thing that directs all your thoughts in the way to attain it."[7]

---

[6] From a medieval hexameter: *Quidquid agis prudenter agas et respice finem.*

[7] *Leviathan,* chapter III.

Looking at the end, we are waiting until the thought of some means emerges. To shorten the waiting time, we should keenly realize the end: *What do you want? What kind of thing do you want? What is the unknown? What is the conclusion?* We should also make a resolute effort to conjure up some appropriate means: *How can you get this kind of thing? Where can you get this kind of thing? In which shop can you buy this kind of thing? How can you find this kind of unknown? How can you derive such a conclusion?*

The last two questions specifically refer to mathematical problems, one to a problem to find, the other to a problem to prove. Let us consider these two cases separately.

(1) *Problems to find.* We consider, as we have done in sect. 9.4, a half-concrete problem: "Find the volume of a pyramid, being given . . . ." The unknown, the volume of a pyramid, is specified, whereas the condition and the data remain unspecified. *How can you find this kind of unknown?* How can we compute the volume of a pyramid? *From which data* can you get this kind of unknown? The proposed problem has data, of course, but the trouble is that at least for the moment we can not derive the unknown from the proposed data. What we really want are *more manageable data;* in fact, we want another, more accessible, *problem with the same unknown.*

If we find such a problem, we may face different situations.

(2) *Problem with the same unknown, formerly solved.* If we are lucky enough to recall such a problem we may proceed to choose *its data as target for auxiliary problems.* This procedure is very often useful. Let us illustrate it by our (previously mentioned, semiconcrete) example.

The unknown of our problem is the volume V of a pyramid. In the most familiar problem with this unknown, the data are B, the area of the base, and h, the length of the altitude. We know the solution of the familiar problem ($V = Bh/3$), and we have recalled it. How can we use this solution? The most natural thing is to try to compute B and h from the data of the proposed (unsolved) problem. In trying this, we choose B and h as our targets; we introduce two auxiliary problems, the unknown of one is B, the unknown of the other is h, and the data in both are the data of our present problem. (For a concrete case, see ex. 4.17, 4.18.)

(3) The foregoing procedure is often applicable, and in many cases it should be *repeatedly* applied.

Let x denote our primary unknown, the unknown of the proposed problem. We are looking for manageable data and we notice that we could find x if we had $y', y'', y''', \ldots$ (by using the solution of a formerly solved problem.) We choose $y', y'', y''', \ldots$ as our new targets, as secondary unknowns. Now, we could find $y', y'', y''', \ldots$ if we had $z', z'', z''', \ldots$ (by using the solutions of several formerly solved problems), and so we set up $z', z'', z''', \ldots$ as targets, as our tertiary unknowns. And so on. We are *working backward,* (see sect. 8.2).

To be well prepared for such work, we should have a store of (simple, often usable) solved problems, and it should be a well-stocked and well-organized store (ex. 12.3).

(4) *Problem with the same unknown, not yet solved.* We may regard such a problem as a stepping stone to the proposed problem, introduce it as an auxiliary

problem, and try to solve it—and such a procedure may be profitable. Yet the prospects are, other things being equal, less favorable than in case (2). In fact, to profit by such a problem in the most obvious way, we should first solve it and then, in addition, we should be able to use it in the manner described under (2).

(5) If we do not see at all how we could find this kind of unknown with which our present problem is blessed, if we cannot recall any formerly solved problem, nor imagine a manageable new one, with the same kind of unknown, we may look for a problem with a *similar kind of unknown.* For instance, if we have to find the volume of a pyramid and we see no other way we may try to recall how we find the area of a triangle and examine various approaches, looking for suggestive analogies.

(6) *Problems to prove.* We could repeat here with little change what we have said about problems to find, but an accelerated survey will be enough.

Also here, it is good to start from a half-concrete example. We have to prove a theorem of the form: "If . . . , then the angle is a right angle." The conclusion of this proposition is specified: "the angle is a right angle" but its hypothesis remains unspecified. *How can you prove such a conclusion? From which hypothesis* can you derive such a conclusion? These questions prompt us to look for a *theorem with the same conclusion* where the statement "the angle is a right angle" is inferred from some other, more manageable hypothesis.

If we are lucky enough to recall a *formerly proved theorem with the same conclusion* we can choose its hypothesis as *target:* we may try to prove the hypothesis of the theorem we have recalled from the hypothesis of the theorem we are trying to establish.

This procedure is often applicable. In many cases we can apply it *repeatedly* and discover the proof for the desired conclusion by *working backward.*

If we come across a theorem with the same conclusion as the proposed theorem but which is equally unproved, we may try to prove it. Such a trial may be profitable, but the prospects should be carefully weighed.

If we cannot recall any formerly proved theorem, nor imagine any manageable new one, with the same conclusion, we may look for a theorem with a *similar conclusion.*

(7) Whatever our problem is, we can be certain in advance that we shall use some formerly acquired knowledge in solving it. Yet, especially if our problem is difficult, we cannot so confidently foretell *which* pieces of knowledge we shall be able to use. Any formerly solved problem, any formerly proved theorem, could be usable, especially if it has some point of contact with our present problem—but we have no time to examine all of them. The foregoing discussion directs our attention to the most likely point of contact. If we have a problem to find, *formerly solved problems with the same kind of unknown*—and if we have a problem to prove, *formerly proved theorems with the same conclusion*—are more likely to be usable. Therefore, we should give high priority to the questions: *How can you find this kind of unknown? How can you prove such a conclusion?*

9.3. *Removing, or adding, a clause.* When our work is progressing slowly or not at all, we become impatient with it and wish we had another problem. Now, it is good to be acquainted with modifications of the problem leading to related problems, the consideration of which has some chance to be useful. Here is a list of the most obvious modifications of this kind.

Problems to find:

(1) Removing a clause from the condition.
(2) Adding a clause to the condition.

Change (1) renders the condition *wider*, (2) renders it *narrower*.

Problems to prove:

(1) Removing an assumption from the hypothesis.
(2) Adding an assumption to the hypothesis.
(3) Removing an assertion from the conclusion.
(4) Adding an assertion to the conclusion.

Both (1) and (4) render the theorem *stronger*, both (2) and (3) render the theorem *weaker*.

The effects of these changes are discussed in exs. 9.4 and 9.5.

9.4. *Widening, or narrowing, the condition.* We consider two conditions, $A(x)$ and $B(x)$, regarding objects x of the same category. We say that $A(x)$ is narrower than $B(x)$, or (which amounts to the same) that $B(x)$ is wider than $A(x)$ iff any object satisfying $A(x)$ necessarily also satisfies $B(x)$. (That is, we use these terms "inclusively;" the case in which $A(x)$ and $B(x)$ have the same scope is included in both terms.)

(1) *Widening the condition* means passing from the proposed problem to another problem which has a wider condition than the proposed. The reader should realize that we have very often performed this operation in the foregoing chapters (of course, without describing it in just these terms.) Thus, in a problem of geometric construction appropriately formulated (or reformulated) the condition refers to a point; we obtain a locus for the point by *keeping only a part of the condition* and removing the other part, that is, by widening the condition. Again, in setting up one equation of a system of equations for several unknowns, we take only one part (requirement, clause, proviso, . . .) of the full condition into account and so, in fact, we widen the condition.

Widening the condition is certainly useful if we can accomplish two things. First, find (describe, list, . . .) the set of all objects satisfying the wider condition. Second, remove from this set those objects that do not satisfy the original condition. I think the reader is aware how these two objectives are attained by the pattern of two loci; he should also review sect. 6.3(3) and some examples and comments dealing with puzzles (see also ex. 6.21).

Yet widening the condition can be useful also in a different manner as the reader familiar with the Cartesian pattern can easily see.

(2) *Narrowing the condition* means passing from the proposed problem to another problem which has a narrower condition than the proposed. We do not have much opportunity to employ this procedure on the level with which we are principally concerned, but here is an example.

We have to solve the equation

$$x^n + a_1 x^{n-1} + a_2 x^{n-2} + \cdots + a_n = 0$$

of higher degree n the coefficients of which, $a_1, a_2, \ldots a_n$, are integers. It is advisable to begin the work by looking for integral roots. In fact, by imposing on x the additional requirement that it should be an integer we narrow the condition. Yet the search for integral roots (they must be divisors of the last coefficient a_n) is comparatively easy, and, if we succeed in finding such a root, we can lower the degree of the proposed equation and so facilitate the search for the remaining roots. (For a concrete case, see ex. 2.31.)

Narrowing the condition is often useful on a more advanced level; see ex. 9.11.

9.5. *Examining a stronger, or a weaker, theorem.* We consider two clearly stated propositions, A and B. If we know that A follows from B (if we can derive A in supposing B true) we say that A is weaker than B and (what means the same) that B is stronger than A. This relation between A and B is particularly interesting when we can neither prove nor disprove A, nor prove or disprove B.

(1) *Examining a possible ground.* We want to prove that two given quantities are unequal. For instance, we want to prove theorem A, which asserts that

$$e < \pi$$

We are lucky enough to observe a third quantity with which both given quantities can be conveniently compared. In our example both e and π are conveniently comparable to the simple number 3. Therefore, to establish A, we consider the theorem B that asserts that

$$e < 3 \quad \text{and} \quad 3 < \pi$$

Of course, A immediately follows from B. The newly introduced proposition B asserts more and so it is stronger than the proposition A, to prove which was our original problem.

Observe that if we have to prove an inequality between two irrational numbers we are almost obliged to proceed as we have in our example; we should discover a rational number which separates the two irrational numbers. In so doing we reduce the original proposition to a stronger proposition, as in our example; the discovery of the separating rational number makes the new proposition stronger.

Such things happen in more advanced research at every turn: to prove a proposed theorem A we have to imagine a stronger theorem B from which A follows but which for some reason is more manageable than A. In proving B, we exhibit a "ground" why A is true. Of course, when we are discovering a theorem B from which A follows, we do not know yet whether we shall be able to prove B, we do not know even whether B is true or not. Thus, for the moment, such a B is not yet a "ground" for the proposed A, just a "possible ground." Still, it may be advisable to examine B, this possible ground for A.

(2) *Examining a consequence.* We want to prove that two quantities are equal. For instance, letting S denote the surface area of a sphere with radius r, we want to prove theorem A that asserts that

$$S = 4\pi r^2$$

It may be advisable to try to prove less to begin with, namely theorem B, which asserts that

$$S \leqq 4\pi r^2$$

(We could possibly prove B in approximating the sphere by circumscribed polyhedra.) At any rate, B obviously follows from A, B is a consequence of A, the theorem B is weaker than the theorem A.

The proof of the weaker theorem B, however, could eventually lead us to the proof of the original theorem A. In fact, the considerations used in proving B could suggest a proof for another weaker theorem, for the opposite inequality

$$S \geqq 4\pi r^2$$

(perhaps by a transition from circumscribed polyhedra to inscribed polyhedra). From the combination of the two weaker theorems, however, the original theorem A would follow.

Such things happen in more advanced research at every turn.

If we are unable to prove a proposed theorem A, we imagine a weaker theorem B which we can prove. Then we may manage to use the weaker theorem B as a spring-board, and with the impetus gained from B we attain A. This can happen even with quite elementary theorems. For instance, we may attain the theorem A which deals with the general case by proving first a weaker theorem B which deals with a particular case, and then use B as a spring-board.

Do you know an example?

9.6. Let m and n denote given positive integers, $m > n$. Compare the following problems:

A. Find the common divisors of m and n.
B. Find the common divisors of n and $m - n$.

What is the logical relation between A and B?
If you are required to do A, do you see some advantage in passing from A to B?
Use the hint to find the common divisors of 437 and 323.

†**9.7.** Compare the following problems:

A. Find the maximum of the function $f(x)$.
B. Find the abscissas x where $f'(x)$, the derivative of $f(x)$, vanishes.

What is the logical relation between A and B?
Do you see some advantage in passing from A to B?

9.8. We consider a triangle and let stand:
O for the center of its circumscribed circle
G for its centroid (center of gravity)
E for the point on the line through O and G (Euler line) for which $2OG = GE$
(G is between O and E.)
Consider the two theorems:

A. The three altitudes of the triangle meet in a point.
B. The three altitudes of the triangle pass through the point E.

What is the logical relation between A and B?
Do you see some advantage in passing from A to B?
Prove B.

† **9.9.** Compare the two following problems (the square roots are taken with the positive sign):

A. Prove that

$$\lim_{x \to +\infty} (\sqrt{x + 1} - \sqrt{x}) = 0$$

B. Given the positive number ϵ, find those positive values of x for which

$$\sqrt{x + 1} - \sqrt{x} < \epsilon$$

What is the logical relation between A and B?
Do you see some advantage in passing from A to B?
Solve B.

9.10. Compare the two following problems (n denotes a positive integer):

A. Prove (or disprove) the proposition: If $2^n - 1$ is a prime number, n must be a prime number.

B. Prove (or disprove) the proposition: If n is a composite number, $2^n - 1$ must be a composite number.

What is the logical relation between A and B?
Do you see some advantage in passing from A to B?
Solve B.

9.11. *The search for a counterexample.* A *counterexample* explodes a statement purporting to apply to all objects of a certain category: The counterexample is an object of the proper category to which the allegedly general statement does *not* apply. The search for counterexamples has several interesting features which we should discuss, although we are obliged to go a little beyond our usual level for a sufficiently instructive illustration.

† (1) *A problem to prove.* Prove, or disprove, the following statement:

If the infinite series with real terms $a_1 + a_2 + a_3 + \cdots$ is convergent, the infinite series $a_1^3 + a_2^3 + a_3^3 + \cdots$ is also convergent.

After more or less work we may surmise that the proposed statement is false and we try to explode it by a counterexample.

† (2) *A problem to find as auxiliary problem to a problem to prove.* We seek a counterexample, that is, an infinite sequence that satisfies the hypothesis, but does *not* satisfy the conclusion, of the statement proposed under (1). We have so, in fact, a problem to find. Let us look at its principal parts.

What is the unknown? An infinite sequence $a_1, a_2, a_3, \ldots,$ of real numbers.
What is the condition? It consists of two clauses:
(I) the series $a_1 + a_2 + a_3 + \cdots$ is convergent
(II) the series $a_1^3 + a_2^3 + a_3^3 + \cdots$ is divergent
We should note that this problem to find arises as an auxiliary problem to a problem to prove.

(3) *Wanted* ONE (anyone) *object fulfilling the condition.* On the elementary level, we are usually required to find all solutions, all the objects satisfying the condition of the problem. Yet in the present case it is enough to find one solu-

tion, one such object; one counterexample is enough to upset the allegedly general statement.

This situation, different from the usual, may demand a different strategy. Leibnitz[8] has some advice to offer: "All solutions may be required or only some solutions. If just any one solution is required, we should invent additional conditions compatible with the original conditions which often demands great skill."

†(4) *Narrowing the condition.* We survey convergent series, satisfying part (I) of the condition, hoping to meet with one that also satisfies part (II). It is natural to begin our search with the simplest and more familiar cases.

Thus we may think first of convergent series with positive terms a_n. Yet, in such a series, $a_n < 1$ for large n, therefore $a_n^3 < a_n$, and so the series with the general term a_n^3 is also convergent; requirement (II) is not fulfilled. We must examine convergent series with positive and negative terms.

The most familiar case here is that of an alternating series in which the signs of the terms form the pattern

$$+ \ - \ + \ - \ + \ - \ + \ - \ + \ - \ \cdots$$

If the terms a_n of such a series steadily decrease to 0 in absolute value, the series is convergent—yet then the terms a_n^3 behave in the same way, form a convergent series and, again, part (II) is not fulfilled. And so we must proceed to less familiar regions.

As we are reluctant to venture too far away from the familiar we may hit on the idea of imposing a restriction:

(III) the signs of the terms a_n should form the pattern

$$+ \ - \ - \ + \ - \ - \ + \ - \ - \ + \ - \ - \ \cdots$$

Even after adding (III) to (I) and (II), we still retain a wide margin of uncertain and arbitrary choice. And so we may hit on the idea of imposing one more (in fact, not quite definite) restriction:

(IV) the series $a_1^3 + a_2^3 + a_3^3 + \cdots$ should diverge in the manner of the familiar series $1 + \frac{1}{2} + \frac{1}{3} + \cdots$.

The self-imposed additional requirements (III) and (IV) essentially *narrow the condition* (see ex. 9.4). They may guide the search for a counterexample, but they may also restrain it. I think that they are more help than hindrance, yet the reader should try to find a counterexample by himself and form his own opinion.

(5) *An alternating procedure.* This may be a good occasion to mention a procedure with which everybody who wants to acquire some ability to do problems to prove should be familiar. (On the high school level there is usually not much opportunity to acquire or practice such ability.)

A problem to prove is concerned with a clearly stated assertion A of which we do not know whether it is true or false: we are in a state of doubt. The aim of the problem is to remove this doubt, to prove A or to disprove it.

Sometimes we are able to devise an approach which could work both ways, which brings us nearer to proof or disproof whatever is in the cards, and so nearer to the solution in any case. Yet such approaches are rare. If we are not lucky

---

[8] *Opuscules et fragments inédits,* p. 166.

enough to find one we face a decision: should we try to prove the assertion A or should we try to disprove it? We have here a choice between two different directions. To prove A we should look for some propositions from which, or for some strategy by which, we could derive A. To disprove A we should look for a counterexample.

A good scheme is to work alternately, now in one direction, then in the other. When the hope to attain the end in one direction fades, or we get tired of working in that direction, we turn to the other direction, prepared to come back if need be, and so, by learning from our work in both directions, we may eventually succeed.

(6) There is a more sophisticated modification of this alternating procedure which may be needed in more difficult cases and may attain a higher aim.

If we cannot prove the proposed assertion A we try to prove instead a weaker proposition (which we have more chances to prove.) And, if we cannot disprove the proposed assertion we try to disprove instead a stronger proposition (which we have more chances to disprove.) If we *succeed in proving* a proposition P we try next to *disprove* a proposition (appropriately chosen) *stronger* than P. Yet if we *succeed in disproving* a proposition P we try next to *prove* a proposition (appropriately chosen) *weaker* than P. Working toward the proposed A from both sides, we may finally prove A. Or we may pass beyond A, and either prove a proposition stronger than A, or we may disprove A yet still save some part of it in proving a proposition weaker than A.

In this way, by working alternately on proofs and counterexamples, we may attain a fuller knowledge of the facts. We may discover a theorem of which we know not only that it is true (we have proved it), but also that it cannot be too easily improved (we have disproved sharper theorems). We catch here a glimpse into the role of proofs in building up science. (*Cf.* G. Pólya and G. Szegö, *Aufgaben and Lehrsätze aus der Analysis,* vol. 1, p. VII. Also MPR, vol. 1, p. 119, ex. 14.)

(7) For further sophistications, striking historical examples, and philosophical overtones of the alternating procedure discussed see the work of I. Lakatos quoted in the Bibliography.

9.12. *Specialization and generalization* are important sources of useful auxiliary problems.

Let us take as example a problem from the theory of numbers. We wish to investigate the *number of divisors* of the positive integer n for which we introduce the symbol $\tau(n)$. For example (we are *specializing*), 12 has the six divisors 1, 2, 3, 4, 6, and 12 and, therefore, $\tau(12) = 6$; we have counted here the "trivial divisors" 1 and 12 of 12, and we intend to proceed the same way for any n.

One way to *specialize* is to consider individual numbers, observing for instance that $\tau(30) = 8$. Or we may systematically list the values of $\tau(n)$ for $n = 1, 2, 3, \ldots$ in constructing a table which starts as follows.

$$\tau(1) = 1 \qquad \tau(6) = 4$$
$$\tau(2) = 2 \qquad \tau(7) = 2$$
$$\tau(3) = 2 \qquad \tau(8) = 4$$
$$\tau(4) = 3 \qquad \tau(9) = 3$$
$$\tau(5) = 2 \qquad \tau(10) = 4$$

Another way to specialize the problem is to consider particular *classes* of numbers. If p is a prime number,

$$\tau(p) = 2, \qquad \tau(p^2) = 3, \qquad \tau(p^3) = 4$$

and now we may find in *generalizing* the answer for any power of p:

$$\tau(p^a) = a + 1$$

If p and q are two different prime numbers, pq has just four divisors 1, p, q, and pq and so

$$\tau(pq) = 4$$

Then we may consider the product of three different primes, and so on. By generalizing, we may attempt to find $\tau(n)$ when $n = p_1 p_2 \ldots p_l$ is the product of l different primes. And so on, sometimes specializing and then again generalizing, we may discover a general expression for $\tau(n)$. (Find it!)

Such are the ways of discovery not only in the theory of numbers (of positive integers) but also in other branches of mathematics and in science in general. In specializing, we try to carve out a more tangible, more accessible part of the problem; by generalizing, we try to extend what we have succeeded in observing in a restricted domain.[9]

9.13. *Analogy* is another fertile source of discovery. In simple cases, we may almost copy the solution of an obviously similar problem. In more delicate cases, a more subtle analogy may not give us immediate material help, but it may indicate the direction in which we should work.

The uses of analogy are of inexhaustible variety; they are illustrated by many examples in the foregoing (and in the following) chapters. Let us quote just one [sect. 1.6(3)]. The problem is to construct an angle in a spherical triangle of which the three sides are given. The construction uses the analogous problem of plane geometry as auxiliary problem: to construct an angle in an ordinary triangle of which the three sides are given.

Remember a few more pairs of analogous problems.

There are, as we have hinted, many other ways to use analogy.[10]

9.14. *And if we fail?* The hope with which we undertake the investigation of an auxiliary problem may be disappointed, our undertaking may fail. Still, the time and effort spent on the auxiliary problem need not be lost; we may learn from failure.

We wish to prove theorem A. We notice a stronger theorem B from which A follows. We undertake to investigate B—if we succeed in proving B, A will also be proved. Yet B turns out to be false. This is disappointing—but our experience with B may lead us to a better evaluation of the prospects of A.

We wish to prove the theorem A. We notice the theorem B, a consequence of A which seems to be more manageable than A. We undertake to investigate B—if we succeed in proving B, we may use it as a stepping stone to prove A.

---

[9] See HSI, Generalization pp. 108–110, Specialization pp. 190–197; and MPR, chapter II and *passim*.

[10] See HSI, Analogy pp. 37–46; and MPR, chapter II and *passim*.

In fact, we manage to prove B, but all our trials to use it as a stepping stone to A fail. This is disappointing—but our experience with B may lead us to a better evaluation of the prospects of A.[11]

9.15. *More problems.* Having observed that an auxiliary problem was helpful in solving some problem, try to understand why it was helpful and where it comes from.

Why? Clarify the relation between the problem and the auxiliary problem; see ex. 9.6–9.10.

Wherefrom? Was the auxiliary problem suggested, or could it have been suggested by regressive work (working backward), generalization, specialization, or analogy? Or was there some other (less usual) source?

[11] *Cf.* MPR, especially vol. 2, pp. 18–20.

CHAPTER 10

THE COMING
OF THE IDEA

*My mind was struck by a flash of lightning in which its desire
was fulfilled.*

DANTE: *Paradiso,* Canto XXXIII.

10.1. Seeing the light

The solution of a problem may occur to us quite abruptly. After brooding over the problem for a long time without apparent progress, we suddenly conceive a bright idea, we see daylight, we have a flash of inspiration. It is like going into an unfamiliar hotel room late at night without knowing even where to switch on the light. You stumble around in a dark room, perceive confused black masses, feel one or the other piece of furniture as you are groping for the switch. Then, having found it, you turn on the light and everything becomes clear. The confused masses become distinct, take familiar shapes, and appear well arranged, well adapted to their obvious purpose.

Such may be the experience of solving a problem; a sudden clarification that brings light, order, connection, and purpose to details which before appeared obscure, confused, scattered, and elusive.

In these matters, however, one grain of experience is worth more than pounds of description. To come closer to personal experience we should get down to a concrete example. Very elementary mathematical examples may be the best to bring us the work, the suspense, and the pleasure of discovery and to "accustom our eyes to see the truth clearly and distinctly." (The last phrase is borrowed from Descartes.)

10.2. Example

I take the liberty of trying a little experiment with the reader. I shall state a simple but not too commonplace theorem of geometry, and then

54

I shall try to reconstruct the sequence of ideas that led to its proof. I shall proceed slowly, very slowly, revealing one clue after the other, and revealing each clue gradually. I think that before I have finished the whole story, the reader will seize the main idea (unless there is some special hampering circumstance). But this main idea is rather unexpected, and so the reader may experience the pleasure of a little discovery.

A. *If three circles having the same radius pass through a point, the circle through their other three points of intersection also has the same radius.*

This is the theorem that we have to prove. The statement is short and clear, but does not show the details distinctly enough. If we *draw a figure* (Fig. 10.1) and *introduce suitable notation,* we arrive at the following more explicit restatement:

B. *Three circles k, l, m have the same radius r and pass through the same point O. Moreover, l and m interesect in the point A, m and k in B, k and l in C. Then the circle e through A, B, C has also the radius r.*

Figure 10.1 exhibits the four circles *k, l, m,* and *e* and their four points of intersection *A, B, C,* and *O.* The figure is apt to be unsatisfactory, however, for it is not simple, and it is still incomplete; something seems to be missing; we failed to take into account something essential, it seems.

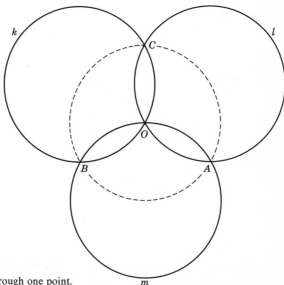

Fig. 10.1. Three circles through one point.

We are dealing with circles. What is a circle? A circle is determined by center and radius; all its points have the same distance, measured by the length of the radius, from the center. We failed to introduce the common radius *r*, and so we failed to *take into account an essential part of the hypothesis.* Let us, therefore, introduce the centers, *K* of *k*, *L* of *l*, and *M* of *m*. Where should we exhibit the radius *r*? There seems to be no reason to treat any one of the three given circles *k*, *l*, and *m* or any one of the three points of intersection *A*, *B*, and *C* better than the others. We are prompted to connect all three centers with all the points of intersection of the respective circle: *K* with *B*, *C*, and *O*, and so forth.

The resulting figure (Fig. 10.2) is disconcertingly crowded. There are so many lines, straight and circular, that we have much trouble in "seeing" the figure satisfactorily; it "will not stand still." It resembles certain drawings in old-fashioned magazines. The drawing is ambiguous on purpose; it presents a certain figure if you look at it in the usual way, but if you turn it to a certain position and look at it in a certain peculiar way, suddenly another figure flashes on you, suggesting some more or less witty comment on the first. Can you recognize in our puzzling figure, overladen with straight lines and circles, a second figure that makes sense?

· ·

We may hit in a flash on the right figure hidden in our overladen drawing, or we may recognize it gradually. We may be led to it by the

Fig. 10.2. Too crowded.

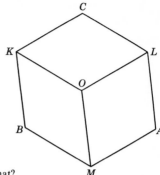

Fig. 10.3. It reminds you—of what?

effort to solve the proposed problem, or by some secondary, unessential circumstance. For instance, when we are about to redraw our unsatisfactory figure, we may observe that the *whole* figure is determined by its *rectilinear part* (Fig. 10.3).

This observation seems to be significant. It certainly simplifies the geometric picture, and it possibly improves the logical situation. It leads us to restate our theorem in the following form.

C. *If the nine segments*

$$K\,O, \qquad K\,C, \qquad K\,B,$$
$$L\,C, \qquad L\,O, \qquad L\,A,$$
$$M\,B, \qquad M\,A, \qquad M\,O,$$

are all equal to r, there exists a point E such that the three segments

$$E\,A, \qquad E\,B, \qquad E\,C$$

are also equal to r.

This statement directs our attention to Fig. 10.3. This figure is attractive; it reminds us of something familiar. (Of what?)

Of course, certain quadrilaterals in Fig. 10.3, such as *OLAM* have, by hypothesis, four equal sides, they are rhombi. A rhombus is a familiar object; having recognized it, we can "see" the figure better. (Of what does the *whole* figure remind us?)

Opposite sides of a rhombus are parallel. Insisting on this remark, we realize that the 9 segments of Fig. 10.3 are of three kinds; segments of the same kind, such as *AL*, *MO*, and *BK*, are parallel to each other. (Of what does the figure remind us *now?*)

We should not forget the conclusion that we are required to attain. Let us assume that the conclusion is true. Introducing into the figure the center *E* of the circle *e*, and its three radii ending in *A*, *B*, and *C*, we

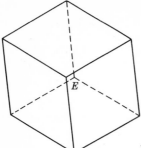

Fig. 10.4. Of course!

obtain (supposedly) still more rhombi, still more parallel segments; see Fig. 10.4. (Of what does the whole figure remind us *now?*)
. . . .
Of course, Fig. 10.4 is the projection of the 12 edges of a parallelepiped having the particularity that the projection of all edges are of equal length.

Figure 10.3 is the projection of a "nontransparent" parallelepiped; we see only 3 faces, 7 vertices, and 9 edges; 3 faces, 1 vertex, and 3 edges are invisible in this figure. Figure 10.3 is just a part of Fig. 10.4, but this part defines the whole figure. If the parallelepiped and the direction of projection are so chosen that the projections of the 9 edges represented in Fig. 10.3 are all equal to *r* (as they should be, by hypothesis), the projections of the 3 remaining edges must be equal to *r*. These 3 lines of length *r* are issued from the projection of the 8th, the invisible vertex, and this projection *E* is the center of a circle passing through the points *A*, *B*, and *C*, the radius of which is *r*.

Our theorem is proved, and proved by a surprising, artistic conception of a plane figure as the projection of a solid.

(The proof uses notions of solid geometry. I hope that this is not a great wrong, but if so it is easily redressed. Now that we can characterize the situation of the center *E* so simply, it is easy to examine the lengths *EA*, *EB*, and *EC* independently of any solid geometry. Yet we shall not insist on this point here.)

10.3. The nature of the helpful idea

The foregoing exemplifies various points about the nature of the helpful idea. Its coming was presented as extremely slow; instead of being uttered triumphantly, it was stuttered. (This was done to give an opportunity to the reader to participate in the discovery.) Also in other respects, our example is somewhat one-sided, as any example is bound to be, because of the immense variety of the phenomenon. Still, if the reader considers it with sympathetic understanding, in the proper

light, in the proper setting, against the background of his own experience, our example may serve as a useful illustration of various
features which occur frequently.

Very often the helpful idea arises *spontaneously*. It brings in some
conspicuous new element, and it *changes our mode of conception.* A
strong *conviction* that the end is within reach follows in its wake.

The *spontaneity* is a very characteristic feature but rather hard to
describe. If it happened to the reader that, from the entanglement of
the lines and letters in Fig. 10.2, the image of the parallelepiped
"jumped" out at him unexpectedly, he will understand better what is
meant. He will also understand, perhaps, to a certain extent, what is
meant by inspiration, how it is possible to interpret the sudden appearance of an impressive idea as the whispering of an inner voice,
or a warning given by a supernatural being.

The *conspicuous new element* arising in our case is the idea of the
parallelepiped. It is rather strange that the appearance of a solid figure
brings the decisive step in the solution of a problem of plane geometry.
It is much more usual that the decisive new element is found within
the domain to which the problem belongs. If the problem is one of
plane geometry we would expect that the new element will be a new
line added to the figure, or a relevant theorem suddenly remembered,
or something of this kind.

The *change in the mode of conception* is very spectacular in the present
case. The circles recede and vanish in the background, the straight
lines come to the foreground. But we cease to conceive them as radii,
we now relate everything to the parallelepiped. The former radii, their
endpoints, the included quadrilaterals obtain a new significance, they
represent now edges, corners, faces of the solid. The change in the
mode of conceiving the elements of the problem is spectacular but
typical. The same kind of revolutionary restructuring of our whole
conception is brought about by the decisive idea in the solution of
almost any other problem. As the idea emerges, the elements assume
new roles, obtain new meanings. In the solution of geometric problems
the elements are reshuffled and regrouped; they are assembled into
triangles, or into pairs of triangles with corresponding sides, or into
rhombi, or into any other familiar configuration serving the purpose
of the search. A line that before the coming of the helpful idea was
just a line, obtains some significance: it becomes the side of a triangle
whose congruence with some other triangle is essential for the solution;
or it becomes a transversal cutting two parallels; or it fits in some other
way into a comprehensive picture. After the coming of the idea we
see more—more meaning, more purpose, and more relations. The
coming of the idea is similar to switching on the light in a dark room.

The helpful idea arises with the *conviction* that the end is within reach. A suddenly arising idea, displaying a spectacular new element amid dramatic rearrangement, has an impressive air of importance and carries strong conviction. This conviction is expressed by such exclamations as "Now I have it!" "I have got it!" "That's the trick!" In our present case, if you just see the parallelepiped without seeing that it leads to the solution, you do not yet have the decisive idea. You need more. You need not see in detail *how* the parallelepiped leads to the solution, but you should have a strong feeling that it *will* lead to the solution.

10.4. Ideas depend on chance

Have you got an idea? You say "Yes"? Then you are lucky. You cannot compel helpful ideas to appear. I take my problem earnestly. I put it to myself. I set it to myself. I realize it keenly. I become absorbed in my problem. I am waiting for a helpful idea; will it come? Perhaps at once, perhaps after some time, perhaps not at all.

We need helpful ideas, we naturally desire to have helpful ideas at our service. But, in fact, they are our masters and they are capricious and self-willed. They may flash upon us unexpectedly, but more often they are long in coming, and sometimes they just keep us waiting and do not turn up at all.

Ideas come when they want to come, not when we want them to come. Waiting for ideas in gambling.

If ideas came quite at random, the solution of problems would depend mainly on chance. Many people believe that this is so. Samuel Butler expressed this opinion in four witty lines:

All the inventions that the world contains,
Were not by reason first found out, nor brains;
But pass for theirs who had the luck to light
Upon them by mistake or oversight.

It is difficult to believe that such a widespread opinion should be entirely devoid of foundation, completely wrong. But is it completely right? Must we depend on chance alone when we have a problem to solve? I hope that, after all the preceding chapters, the reader has an opinion.

Examples and Comments on Chapter 10

10.1. *The spontaneity of ideas. A quotation and a comment.*
(1) We quote from *The Age of Reason,* First Part, by Thomas Paine.

Any person who has made observations on the state and progress of the human mind, by observing his own, cannot but have observed that there are two distinct classes of what are called thoughts: those that we produce in ourselves by reflection and the act of thinking, and those that bolt into the mind of their own accord. I have always made it a rule to treat those voluntary visitors with civility, taking care to examine, as well as I was able, if they were worth entertaining; and it is from them I have acquired almost all the knowledge that I have.

(2) Lichtenberg observed that we should not say "I am thinking" but rather "it is thinking" as we say "it is raining" or "it thunders." Lichtenberg suggests that there are spontaneous actions of the mind which we cannot command any more than the great forces of nature.

We could also say that our mind sometimes behaves like a sort of horse or mule, some strange animal whom we have to humor and occasionally to cudgel in order to get the desired service from it—which we quite often do not get.

(Georg Christoph Lichtenberg, 1742–1799, German physicist and writer; the *Aphorismen* may be his best remembered work.)

10.2. *Two experiments.* Some time (but not too much time) spent on crossword puzzles may be quite rewarding; we can learn something about problem solving, how we think, and how we ought to think.

(1) In a crossword puzzle you find the following clue: "A very common kind of heart (10 letters)." You may have no inkling at the outset what the word is or what the clue means. Yet a crossing word that you have succeeded in finding in the meantime yields some information: a letter in the middle. Another crossing word supplies a second letter; then you find a third letter, or a fourth—and suddenly the desired word "bolts into your mind."

Provide yourself with a sheet of paper and turn to the solution of this problem on p. 172. First cover the whole solution with your sheet. Then, sliding it downward, uncover just one line—do you know now the desired word? If not, uncover a second line, then the next, and so on; you may experience the idea "bolting into your mind."

†(2) If you know just a little calculus (not too much) you may try a similar experiment in finding the value of an indefinite integral. Take a sheet and turn to p. 172.

CHAPTER 11

THE WORKING
OF THE MIND

*Mariotte says that the human mind is like a bag: when you are
thinking you are shaking the bag until something falls out of it. Hence
there is no doubt that the result of thinking depends to some extent
on chance. I would add that the human mind is more like a sieve:
when you are thinking you are shaking the sieve until some minute
things pass it. When they pass, the spying attention catches whatever
seems relevant. Again, it is something like this: to catch a thief, the
commander of a city orders the whole population to pass a certain gate
where the man who was robbed is watching. Yet, to save time and
trouble, some method of exclusion may be used. If the man robbed
says that the thief was a man, not a woman, and an adult, not a
youngster or a child, those not concerned are excused from passing
the gate.*

LEIBNITZ: *Opuscules et fragments,* p. 170.

11.1.   How we think

A problem solver must know his mind and an athlete must know
his body in about the same way as a jockey knows his horses. I imagine
that a jockey studies horses not for the sake of pure science but to make
them perform better, and that he studies more the habits and whims
of individual horses than horse physiology or horse psychology in
general.

What you start reading now is not a chapter in a textbook of
psychology; it is not exactly a conversation between problem solvers
who talk about the habits of their minds as jockeys may talk about the
habits of their horses; it is, however, more like a conversation than a
formal presentation.

11.2. Having a problem

An essential ingredient of the problem is the desire, the will, and the resolution to solve it. A problem that you are supposed to do and which you have quite well understood, is not yet your problem. It becomes your problem, you really have it, when you decide to do it, when you desire to solve it.

You may be involved more or less deeply in your problem—your desire to solve it may be more or less strong. Unless you have a very strong desire, your chances to solve a really hard problem are negligible.

The desire to solve your problem is a *productive* desire: it may eventually produce the solution, it certainly produces a change in your mental behavior.

11.3. Relevancy

You may have a problem so badly that the problem has you; you cannot get rid of your problem, it follows you everywhere.

A man with a problem may be obsessed by his problem. He appears absentminded; he does not notice things which appear obvious to his neighbors, and he forgets things which none of his neighbors would forget. Newton, working intensely on his problems, often forgot to eat his meals.

Yes, the problem solver's attention is *selective:* it refuses to dwell on things which appear irrelevant to his problem and espies the most minute things that appear relevant. It is a "spying" attention as Leibnitz put it.

11.4. Proximity

A student takes a written examination in mathematics. He is not required to do all the proposed problems, but he should do as many as possible. In this situation his best strategy may be to start by looking through all the problems at an appropriate pace and choose those he is most likely to master.

Observe that this supposes that the problem solver is able to assess to some extent the difficulty of his problems, that he can estimate to a degree his "psychological distance" from his problem's solution. In fact, anybody seriously concerned with his problem has a vivid feeling for the proximity of the solution and for the pace of his progress toward the solution. He may not use words but he feels keenly: "It goes well, the solution may be just around the corner," or "It goes so slowly and the solution is still far off," or "I got stuck, there is no progress at all," or "I am drifting away from the solution."

11.5. Prevision

As soon as we are seriously concerned with our problem, we try to foresee, we try to guess; we expect something, we anticipate an outline of the shape of the solution. This outline may be more or less definite—and of course it may be more or less wrong, although I would say not often very wrong.

All problem solvers guess, but the sophisticated and the unsophisticated guess somewhat differently.

A primitive person just sits there with his problem, scratching his head or chewing his pencil, waiting for a bright idea, and doing little or nothing to bring that bright idea nearer. And when the desired idea eventually appears and brings a plausible guess, he simply accepts that guess, regarding it as the solution with little or no criticism.

A more sophisticated problem solver takes his guesses more skeptically. His first guess may be: "There are 25" or "I should tell him this and that." Yet then he checks his guess and may change it: "No, not 25. Yet let me try 30" or "No. It is no use to tell him that, because he could answer thus and so. Yet I could tell him that . . . ." And eventually, by "trial and error," by successive approximations, the problem solver may arrive at the right answer, at an appropriate plan.[1]

A still more sophisticated and more experienced problem solver, when he does not succeed in guessing the whole answer, tries to guess some part of the answer, some feature of the solution, some approach to the solution, or some feature of an approach to the solution. Then he seeks to expand his guess, but also seeks opportunities to check his guess, and so he seeks to adapt his guess to the best information he can get at the moment.

Of course, both the sophisticated and the unsophisticated would like to have a really good guess, a bright idea.

And everybody would like to know what chances his guess has to come true. Such chances cannot be precisely evaluated (this is not the place to discuss remote possibilities of evaluation). Many times, however, the problem solver has a definite feeling about the prospects of his guess. Primitive people who do not even know what a proof is may have the strongest feelings about their guesses; sophisticated people may distinguish fine shades of feeling; but anybody who has conceived a guess has some feeling about the likely fate of his guess. And so we notice still another sort of feeling, besides the feelings of relevancy and proximity, in the problem solver's mind.

Is this point relevant? How far off is the solution? How good is this guess? Such questions accompany each move of the problem

---

[1] Sect. 2.2(1) and 2.2(5).

solver; they are more felt than formulated and the answers, too, are more felt than formulated. Do such feelings guide the problem solver or do they merely accompany his decisions? Are they causes or symptoms? I don't know, but I do know that if you do not have such feelings, you are not really concerned with your problem.

11.6. Region of search

I seldom part with my wrist watch, but when I do I usually have some trouble to find it. When I miss my watch, I habitually start looking for it at some well-defined place: on my desk, or on a certain shelf where I am used to store little belongings, or at any third place if I happen to remember that I took off my wrist watch just there.

Such behavior is typical. As soon as we are seriously concerned with our problem, we anticipate an outline of its solution. This outline may be vague, it may be hardly conscious, but it manifests itself in our behavior. We may try various solutions, but they are all alike; they are all within that preconceived, but perhaps not consciously preconceived, outline. When none of the solutions tried fits the problem, we feel lost, nothing else comes to mind; we cannot step outside that preconceived outline. We do not look for just any kind of solution, but for a certain kind, a kind within a limited outline. We do not look for a solution just anywhere in the world, but for a solution within a certain limited *region of search*.[2]

To begin our search within a likely limited region may be reasonable. When I am trying to find my missing wrist watch, it is quite reasonable not to look for it anywhere in the universe, or anywhere in the city, or anywhere in the house, but just on my desk where I found it several times in the past. It is quite reasonable to begin by seeking the unknown within that limited region, but it is unreasonable to persevere in seeking it there even when it becomes more and more clear that it is not there.

11.7. Decisions

Problem solving may be contemplative; with primitive people, it may be inarticulate brooding. Or it may be a long, strenuous, winding road to the solution, each turning of which is marked by a decision. Such decisions are prompted (or perhaps merely accompanied) by feelings of relevancy and proximity, by swelling or fading hope. Decisions and prompting feelings are seldom expressed in words, but may be occasionally:

[2] Karl Duncker, On Problem Solving, *Psychological Monographs*, vol. 58, No. 5 (1945). See p. 75.

"Now, let me look at this."

"No, there is not much to see here. Let me look at that."

"There is not much to see here either, but there is something in the air. Let me look at it a little longer."

An important type of decision is to enlarge the region of search, to discard a limitation the narrowness of which starts giving us an oppressive feeling.

11.8. Mobilization and organization

The problem solver's mental activity is very imperfectly known and its complexity may be unfathomable. Yet one result of this activity is perfectly obvious: as the problem solver advances, he collects more and more material.

Let us compare the problem solver's conception of a mathematical problem at the beginning and at the end of his work. When the problem arises, there is a simple picture: the problem solver sees his problem as an undivided whole without details, or with very few details; for instance, he may see just the principal parts, unknown, data, and condition, or hypothesis and conclusion. Yet the final picture is very different: it is complex, full of added details and materials the relevancy of which the problem solver could hardly have suspected at the outset. There are auxiliary lines in the originally almost empty geometric figure, there are auxiliary unknowns, there are materials from the formerly acquired knowledge of the problem solver, especially theorems applied to the problem. That just these theorems will be applicable, the problem solver did not foresee at all at the beginning.

Where do all these materials, auxiliary elements, theorems, etc. come from? The problem solver has collected them; he had to extract them from his memory and purposefully connect them with his problem. We call such collecting *mobilization* and such connecting *organization*.[3]

Solving a problem is similar to building a house. We must collect the right material, but collecting the material is not enough; a heap of stones is not yet a house. To construct the house or the solution, we must put together the parts and organize them to a purposeful whole.

Mobilization and organization cannot actually be separated; they are complementary aspects of the same complex process—of our work aimed at the solution. Such work, when intensive, brings into play all our psychical resources, requires the whole gamut of our mental activities, and presents an inexhaustible variety of aspects. We may be tempted to distinguish some of the manifold mental operations involved and describe them by such terms as isolation and combination, recognizing and remembering, regrouping and supplementing.

[3] *Cf.* ex. 2.74.

The following lines attempt to describe these activities. Of course, the reader should not expect, and could not reasonably expect, hard and fast distinctions or rigid and exhaustive definitions.

11.9. Recognizing and remembering

In examining our problem we cheer up when we *recognize* some familiar feature. Thus, in examining a geometric figure, we may recognize with pleasure a triangle not noticed before, or a pair of similar triangles, or some other intimately known configuration. Examining an algebraic formula we may recognize a complete square, or some other familiar combination. Of course, we may also recognize, and it may be very useful to recognize, some more complex situation to which we cannot yet attach a name and for which we have not yet a formal definition, but which strikes us as familiar and important.

We have good reasons to be pleased when we have recognized a triangle in the proposed figure. In fact, we know several theorems and have solved various problems about triangles, and one or the other of these known theorems or former solutions may be applicable to our present problem. By recognizing a triangle, we establish contact with an extensive layer of our formerly acquired knowledge, some streak of which might be useful now. Thus, in general, recognizing may lead us to *remembering* something helpful, to mobilization of relevant knowledge.

11.10. Supplementing and regrouping

We have recognized a triangle in the figure and have succeeded in remembering a theorem about triangles that has some chance to be applicable to the present situation. Yet, to actually apply that theorem we must add some auxiliary line to our triangle, for instance, an altitude. Thus, in general, prospective useful elements just mobilized may be added to our conception of the problem to enrich it, to make it fuller, to fill in gaps, to supply its deficiencies, in a word, to *supplement* it.

Supplementing introduces new materials into our conception of the problem and is an important step in its organization. Yet sometimes we can make an important advance in organization without introducing any new material, just by changing the disposition of the elements already present, by conceiving them in new relations, by rearranging or *regrouping* them. By regrouping its elements, we change the "structure" of our problem's conception. Thus regrouping means *restructuring*.[4]

Let us state this consideration more concretely. The decisive step in the solution of a geometric problem may be the introduction of an

4 *Cf.* Duncker, *loc. cit.* pp. 29–30.

appropriate auxiliary line. Yet sometimes we can take the decisive step without introducing any new line, just by conceiving the lines already present in a new fashion. For example, we may notice that certain lines form a pair of similar triangles. In noticing this familiar configuration, we recognize hitherto unobserved relations between the elements of the figure, we see the elements differently grouped, we see a new structure, we see the figure as a better arranged, more harmonious, more promising whole—we have restructured the problem material.

Regrouping may involve a change in emphasis. Elements and relations which were in the foreground before the regroupment may now surrender their privileged place and recede into the background; they may even recede so far that they practically drop out from the conception of the problem. For better organization we must now and then reject things which we thought relevant some time ago. Yet, on the whole, we add more than we reject.

11.11. Isolation and combination

When we are examining a complex whole, our attention may be attracted now by this detail and then by another. We concentrate on a certain detail, we focus on it, we emphasize it, we single it out, we distinguish it from its surroundings, in one word, we *isolate* it. Then the spotlight shifts to another detail, we isolate still another detail, and so on.

After examining various details and revaluing some of them, we may feel the need of visualizing again the situation as a whole. In fact, after the revaluation of some details, the appearance of the whole, the "vue d'ensemble," the "Gestalt" may have changed. The combined effect of our reassessment of certain details may result in a new mental picture of the whole situation, in a new, more harmonious *combination* of all the details.

Isolation and combination may advance the solution in complementing each other. Isolation leads to decomposing the whole into its parts, a subsequent combination reassembles the parts into a more or less different whole. Decomposed and recombined, again decomposed and again recombined, our view of the problem may evolve toward a more promising picture.

11.12. A diagram

A diagrammatic summary of the foregoing sections is offered by Fig. 11.1, which the reader should take for what it is worth. Nine terms are arranged in a square; one occupies the center of the square, four

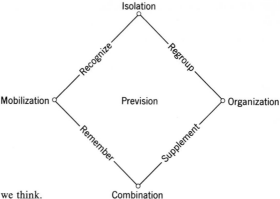

Fig. 11.1. How we think.

others the four vertices, and four more terms are written along the sides.

Mobilization and *organization* are represented by opposite ends of the horizontal diagonal of the square. In fact, these are complementary activities. Mobilization is extracting relevant items from our memory, organization is connecting such items purposefully.

Isolation and *combination* are represented by opposite ends of the vertical diagonal. In fact, these are complementary activities. Isolation is selecting a particular detail from the surrounding whole, combination is assembling dispersed details to a meaningful whole.

The sides adjoining the corner assigned to mobilization are labeled *recognize* and *remember.* In fact, mobilization of items relevant to the problem often starts from recognizing some element given with the problem and consists in remembering connected elements.

The sides adjoining the corner assigned to organization are labeled *supplement* and *regroup.* In fact, organization means supplementing the conception of the problem, making it fuller by adding new details and filling the gaps; and it also means regrouping the whole conception.

As we read the terms along the sides of the square, from left to right, we proceed from mobilized details to the organized whole; a detail just recognized, carefully isolated and focused, may induce a regroupment of the whole conception. Also, a detail remembered which fits into a combination is suitably added to the conception and supplements the whole.

Prevision is the center of our activity aimed at the solution, as the corresponding point is the center of our symbolic square. We keep on mobilizing and organizing, isolating and recombining, recognizing and remembering all sorts of elements, regrouping and supplementing our conception of the problem, just to foresee the solution, or some feature of the solution, or a bit of the path leading to it. If prevision comes to us abruptly, in a flash, we call it inspiration, or illuminating idea; our central desire is to have such an idea.

The mental operations surveyed in Fig. 11.1 take more specific forms when applied to special material. Thus, correspondingly to the four sides of the square, we list four mental operations important in solving mathematical problems:

Recognize:	Regroup:
use definitions	transform the problem
Remember:	Supplement:
known theorems and problems	introduce auxiliary elements

There is another point. The problem solver's moves are prompted or accompanied by feelings of relevancy and proximity, and feelings gauging the goodness of his guess. In discussing this we have mentioned incidentally that more sophisticated people have more differentiated feelings concerning such points. I do not wish to suppress here a rather speculative remark:[5] some such shades of feeling may be connected with the mental operations surveyed in Fig. 11.1.

We cheer up when our conception of the problem appears *well balanced* and *coherent, complete* with all details, and all details are *familiar.* If we have *distinct details* in a *harmonious whole,* the idea of the solution appears *near.* What we express with these terms is, it seems to me, that certain *activities considered above are well progressing,* or have already reached their goal.

Our conception of the problem appears well balanced when we do not feel the need of *regrouping* it, and appears as coherent when we have no trouble in *remembering* its details, but any detail easily recalls the others. When there is no need of *supplementing* it, the conception appears as complete, and it appears as familiar when all details have been *recognized.* Distinctness of details comes from foregoing *isolation* of, and concentration on, each detail, and the harmony of the whole conception results from successful *combination* of the details. We say that the idea is near when we feel that we are well progressing toward fuller *prevision.*

Wishing to arrange these favorable signs of our progress systematically, we place them so that their relative positions are the same as those of the corresponding terms in the square of Fig. 11.1. Thus, we arrange seven terms so as the four sides of that square and the three important points on its vertical diagonal are disposed. See the scheme:

[5] *Cf.* HSI, Signs of progress 4, p. 184.

Well isolated:
distinct details

Well recognized: Well grouped:
familiar   well balanced

Prevision promising:
idea near

Well remembered: Well supplemented:
coherent   complete

Well combined:
harmonious whole

11.13. The part suggests the whole

A whistling boy passed me in the street and I caught one or two measures of a melody which I like very much but had not heard for a long time. Suddenly that melody filled by mind, ousting completely whatever worries or idle thoughts I had before.

This little event is a good illustration of the "association of ideas," a phenomenon already described by Aristotle and by many authors after him. Bradley gives a good description: "Any part of a single state of mind tends, if reproduced, to reinstate the remainder." In fact, in my case, one measure brought back the whole impact of that melody and then, by and by, the remaining measures. Here is another description which lacks essential details but is easy to remember: "The part suggests the whole." Let us regard this short sentence as a convenient abbreviation of Bradley's more precise formulation.

Notice the important words "tends" and "suggests." The statements
"The part suggests the whole"
"The part tends to reinstate the whole"
"The part has a chance to reinstate the whole"
may be acceptable, but the sentence
"The part reinstates the whole"
is certainly inacceptable as an expression for the "law of association": there is no necessity of recall, just a chance, a tendency. We also know something about the strength of that tendency; a part more in the focus of attention suggests the whole stronger; several parts jointly suggest the whole stronger than any one of them singly. These additions are important if we wish to understand the role of association in the problem solver's mental experience.

Let us consider a strongly schematized example. A mathematical problem can be quickly solved by the application of a certain decisive theorem D, but it is very difficult to solve it without D. At the outset, the problem solver does not even suspect that the theorem D is relevant to his problem, although he is quite well acquainted with the theorem D itself. How can he discover the decisive role of D? There are various cases.

The case is relatively simple if the proposed problem and the theorem D have a common component part. The problem solver, after having tried this and that, will come upon that component part, isolate and focus it, and then that common part has a chance to recall or "reinstate" the whole theorem D.

The case is less simple if the original conception of the problem and the decisive theorem D have no common component. Still, if there is another theorem C, also known to the problem solver, that has some component in common with the problem and another component in common with D, the problem solver may attain D by first contacting C and then passing from C to D.

Of course, the chain of associations may be still longer; the proposed problem may be in associative contact with A, A with B, B with C, and finally C with D. The longer the chain the longer must the problem solver "shake the bag" or "shake the sieve" till the decisive D eventually falls out.

Shaking the bag or the sieve is a metaphorical way to describe the problem solver's mental experience (see the quotation prefixed to this chapter). The foregoing sections summarized by Fig. 11.1 attempted to describe this experience somewhat less metaphorically. There is a quite plausible interpretation of the activities described; through them, the problem solver seeks to establish desirable associative contacts.

In fact, in recognizing an element, the problem solver places it in a context with which it has strong associative contact. Any newly mobilized element, added to the problem's conception, offers chances to attract further elements with which it is in associative contact. When the problem solver isolates and focuses an element, the attention spent on it gives it more chance to bring in associated elements. A regroupment may bring together elements which could exercise more associative attraction jointly than anyone could singly.

It is, however, hardly possible to explain the problem solver's mental experience by association *alone;* there must be something else besides associative attraction to distinguish between relevant and irrelevant, desirable and undesirable, useful and useless associated elements and combinations.[6]

Examples and Comments on Chapter 11

11.1. *Your experience, your judgment.* The aim of this book is to improve your working habits. In fact, however, only you yourself can improve your own habits. You should find out the difference between what you are usually doing and what

---

[6] *Cf.* Duncker, *loc. cit.* p. 18.

you ought to do. This chapter was written to help you to see better what you are usually doing.

The following exercises, ex. 11.2–11.6, ask you to illustrate passages of the foregoing text. In the first place, try to find illustrations from your own work—such illustrations as come to your mind spontaneously have the best chance to be illuminating. Try to judge with an open mind whether the descriptions in the text or the illustrations in the solutions agree with your experience.

11.2. *Mobilization.* Recall your work on some problem of geometry where the figure, originally almost empty, became more and more filled by auxiliary elements as the solution progressed.

11.3. *Prevision.* Can you recall a case in which, at a pretty definite moment, you became suddenly convinced that the solution will succeed?

11.4. *More parts suggest the whole stronger.* Can you agree, judging by your own experience?

11.5. *Recognizing.* Can you recall a case in which recognizing an element (noticing its formerly unnoticed familiar role) appeared as the turning point of the solution?

11.6. *Regrouping.* Can you recall a case in which regrouping the figure appeared as the key to the solution?

11.7. *Working from inside, working from outside.* Establishing contacts between the proposed problem and his previous experience is certainly an essential part of the problem solver's performance. He can try to discover such contacts "from inside" or "from outside." He may remain within the problem, examining its elements till he finds one that is capable of attracting some usable element from outside, that is, from his previously acquired knowledge. Or he may go outside the problem, examining his previously acquired knowledge until he finds some element applicable to his problem. Working from inside, the problem solver scans his problem, its component parts, its aspects. Working from outside, he surveys his existing knowledge, and ransacks the provinces of knowledge that are most likely to be applicable to the present problem. The two parts of Fig. 11.2 attempt to give visual expression to "inside" and "outside" work.

Fig. 11.2. Working from inside, working from outside—to pierce the clouds.

11.8. *Heuristic rat's maze.* Figure 11.3 may represent trails in wooded hilly country made by woodcutters without much care, and then abandoned; the point *E* marks the entrance. Figure 11.3 may also represent a maze in which rats are made to run in a psychological experiment.

But Fig. 11.3 may also symbolize the problem solver's activity when he is working in a certain way. After a straightforward beginning, he follows a curving trail until he reaches an (actual or imagined) impasse. Then he turns back and starts retracing his steps, but noticing a side path he follows it to another impasse which makes him turn back. And so he goes on, attempting several trails, retracing many of his steps, noticing new issues, exploring his problem and proceeding on the whole, we hope, in the right direction.

11.9. *Progress.* As the solution progresses, the problem solver's conception of the problem continually changes; especially, he collects more and more material connected with the problem. Let us tentatively assume that we can

Fig. 11.3. Trials, trails, impasses, and side issues.

somehow measure the extent of the material collected at each instant, and that this extent is somehow proportional to the appropriateness of the problem solver's conception of his problem. Of course, this assumption is crude and unlikely, but it enables us to represent graphically the progress of the solution.

In a coordinate system, see Fig. 11.4, we take the time as abscissa and the "measure" of the problem's conception at the instant considered as the corresponding ordinate. The resulting curve represents the problem's conception as function of the time; it gives visual expression to the solution's growth in the problem solver's mind.

Let us assume that the solution develops without setbacks; correspondingly, Fig. 11.4 displays a rising curve, representing a never decreasing function of the time. The curve begins (from the left) with a slowly, almost uniformly rising dotted line, which intends to symbolize the "prehistory," the subconscious growth of the problem. The point *C*, from which the curve is traced in full, marks the beginning of *conscious* work. The slope of the curve at any point represents the pace of progress at the corresponding instant. The pace varies; it is slowest at the point *S*, which is a point of momentary *stagnation;* the tangent to the curve at the point *S* is horizontal. The pace is fastest at the point *I* where the slope is a maximum; *I*, a point of inflection, marks the emergence of the decisive idea, the instant of *inspiration*. (The point *S* is also a point of inflection, but of the opposite kind; the slope at *S* is a minimum.)

The development of the solution in the problem solver's mind is a complex process which has an inexhaustible variety of aspects. Figure 11.4 does not pretend to exhaust these aspects, but it may add a little to our other discussions; it may complement, for instance, the suggestions of Fig. 7.8.

11.10. *You too.* Much of what I know, or imagine to know, about problem solving came to me through reflecting on relatively few suggestive experiences. In reading a book, having a discussion with a friend, talking with a student, or

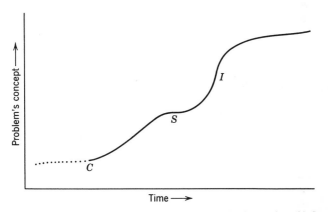

Fig. 11.4. Conscious beginning—stagnation—idea, inspiration, point of inflection.

observing the faces in an audience I suddenly recognized something and was tempted to say to myself: "You too, you are acting very much the same way." In fact I sometimes had this feeling "you too" in observing animals: dogs, birds, and once a mouse.

11.11. *Mice and men.* The landlady hurried into the backyard, put the mousetrap on the ground (it was an oldfashioned trap, a cage with a trapdoor) and called to her daughter to fetch the cat. The mouse in the trap seemed to understand the gist of these proceedings; he raced frantically in his cage, threw himself violently against the bars, now on this side and then on the other, and in the last moment he succeeded in squeezing himself through and disappeared in the neighbour's field. There must have been on that side one slightly wider opening between the bars of the mousetrap. The landlady looked disappointed, and so did the cat who arrived too late. My sympathy from the beginning was with the mouse, and so I found it difficult to say something polite to the landlady, or to the cat; but I silently congratulated the mouse. He solved a great problem, and gave a great example.

That is the way to solve problems. We must try and try again until eventually we recognize the slight difference between the various openings on which everything depends. We must vary our trials so that we may explore all sides of the problem. Indeed, we cannot know in advance on which side is the only practicable opening where we can squeeze through.

The fundamental method of mice and men is the same; to try, try again, and to *vary the trials* so that we do not miss the few favorable possibilities. It is true that men are usually better in solving problems than mice. A man need not throw himself bodily against the obstacle, he can do so mentally; a man can vary his trials more and learn more from the failure of his trials than a mouse.

CHAPTER 12

THE DISCIPLINE
OF THE MIND

*Method consists entirely in properly ordering and arranging the
things to which we should pay attention.*
DESCARTES: *Œuvres,* vol. X, p. 379; Rules for the Direction of the
Mind, Rule V.

12.1. How we ought to think

Chapter 11 attempted to describe the typical mental behavior of
the problem solver. Yet, is the typical also rational? We may behave
so, but should we behave so?

These questions, in their vague generality, can hardly be answered,
but they serve to indicate the trend of this chapter. Guided by the
mental experience of the problem solver, which we surveyed in chapter
11, we shall attempt to list such mental operations (steps, procedures,
etc.) as are typically useful for solving problems and, in listing them,
we shall try to assign the proper position to each in the problem solving
process.

We shall express the typically useful operations of problem solving
in a concise, condensed form by "stereotyped" questions and recommendations. The reader should realize that these questions and
recommendations can be interpreted in two different ways: either as
quotations from a soliloquy of the problem solver, or as addressed by
an understanding teacher to a student of his whom he wishes to help.[1]

---

[1] The reader is urged to compare the exposition of the present chapter with the parallel
passages of HSI; for the foregoing see pp. 1–5, Purpose. The "stereotyped" questions and
recommendations, which I consider as an essential part of my method, were introduced in
my paper no. 2, see the Bibliography.

12.2. Focusing the aim

When you have a problem it often comes to your mind, and it may do so so often that it becomes an obsession. Yet you should not only think of your problem in some vague way, you should face it, you should see it clearly, you should ask yourself: *What do you want?*

There are many opportunities to ask this question in the course of the solution. When you become too deeply engaged in some side issue which may be irrelevant after all, when your thoughts start wandering, it may be advisable to ask yourself: *What do you want?* and bring your aim again into focus.

The aim of a problem to find is the unknown; to focus this aim, you should ask: *What is the unknown?* The aim of a problem to prove is the conclusion, and so the appropriate form of the question is: *What is the conclusion?*

Having clearly seen your aim, the thing you would like to have, you should take stock of the things in your possession which you could possibly use to attain your aim, and so you should ask yourself: *What have you?*

In fact, if you wish to join two points, to find your way from one point to the other, it may help you to look at the points alternately, first at one, then at the other, and so you often have the opportunity to ask the questions in succession: *What do you want? What have you got?*

Adapted to problems to find, the questions are: *What is the unknown? What are the data? What is the condition?* Adapted to problems to prove: *What is the conclusion? What is the hypothesis?*

What do these questions prompt us to do? To turn our attention to the points mentioned. According to Descartes (see the motto of this chapter), method consists in paying attention to all relevant points, one after the other, in appropriate sequence. Now, there is little doubt that the principal parts of a problem to find (unknown, data, and condition) and those of a problem to prove (conclusion and hypothesis) are relevant. They are even so important that it seems appropriate to consider them early; when you have well understood the problem as a whole, pay attention to its principal parts.

12.3. Assessing the prospects

A problem solver earnestly concerned with his problem keenly feels the proximity of the goal and the pace of his progress toward the goal; he keenly feels any change affecting the prospects of his plan. Now and then it is desirable to go a little beyond mere feelings, to see more clearly our position in relation to the problem, to diagnose the problem, to assess the prospects, and such is the tendency of the following questions.

Some problems are hopeless. If the problem in hand is hopeless, we should not get too deeply involved in it, and so we ask: *Is there any answer to this question?* Is there a clear answer, a sensible answer? *If there is an answer, can I find it?*

When we are dealing with a problem to find, we should ask: *Is there a solution?* We may ask more elaborately: *Is there just one solution, or several solutions, or no solution?* Is the condition just enough to determine the unknown, or does it demand too little, or too much?

When we are dealing with a problem to prove, the appropriate question is: *Is the theorem true or false?* We may ask more elaborately: *Is the theorem true? Or is a stronger hypothesis needed to imply the conclusion? Is the theorem sharp? Or is a weaker hypothesis enough to imply the conclusion?* And so on.

In fact, we cannot give a definitive answer to any of these questions before we have finished our work and solved the problem. Yet the questions do not really aim at a definitive answer, they demand only a provisional answer, a guess. In trying to guess right we may *clarify our position* in relation to the problem, and this is the desirable effect. We have stated the foregoing questions in a short, colloquial form. More cautiously, we should ask: Is it *likely* that there is an answer, a solution? The theorem stated may be true or false: what is *more likely?*

How soon should we ask the foregoing questions? There is (and there should be) no hard and fast rule. Very often they naturally follow the questions of sect. 12.2 concerned with the principal parts of the problem.

12.4. Wanted: an approach

The end suggests the means; the consideration of the aim (of the unknown, of the conclusion) may suggest an approach. One question leads to another: *What do you want? What is the unknown? How can you find this kind of unknown? From what data can you derive this kind of unknown?* And these questions may initiate a regressive approach; if we perceive "data" from which the unknown of the proposed problem could be derived, we may choose them as target for an auxiliary problem and so we may start working backward [see ex. 8.1(2)].

For a problem to prove, the corresponding questions are: *What do you want? What is the conclusion? How can you derive this kind of conclusion? From what hypothesis can you derive this kind of conclusion?*

Instead of emphasizing the unknown (the conclusion) we may emphasize the data (the hypothesis): *What are the data? What are such data good for? What can you derive from such data?* For problems to prove there is a corresponding string of questions: *What is the hypothesis? What is such a hypothesis good for? What can you derive from such a hypothesis?* These questions may initiate a progressive approach. (See ex. 8.1; we have discussed there, and we should remember, that, on the whole, regressive planning is preferable to progressive planning.)

Unfortunately, it can easily happen that we cannot design a serviceable plan, neither by regressive nor by progressive work. There are, however, other questions that could suggest an approach, and here are some that we may ask with advantage early in the game: *What kind of problem is it? Is it related to any known problem? Is it like any known problem?* Trying to classify our problem, trying to find relations and resemblances to known problems, we may perceive a familiar pattern applicable to our problem, and then we have something to begin with— we see the first stretch of a road that may lead us to the solution.

Trying to find a usable related problem we may survey those relations that are most often useful: *Do you know a related problem? Could you imagine a related problem? Do you know, or could you imagine, a problem of the same kind, or an analogous problem, or a more general, or a more special problem?* These questions, however, may lead us far from the proposed problem, so it is usually more profitable to ask them a little later when the problem is clarified and well fixed in our mind so that we do not risk losing sight of it when working at some distance from it.

12.5. Wanted: a more promising aspect

When you are handling material things (for instance, when you are about to saw a limb off a tree) you automatically put yourself in the most convenient position. You should act similarly when you are facing any kind of problem; you should try to put yourself in such a position that you can tackle the problem from the most accessible side. You turn the problem over and over in your mind; try to turn it so that it appears simpler. The aspect of the problem that you are facing at this moment may not be the most favorable: Is the problem as simply, as clearly, as suggestively expressed as possible? *Could you restate the problem?*

Of course you want to restate the problem (transform it into an equivalent problem) so that it becomes more familiar, more attractive, more accessible, more promising.

The aim of your work is to bridge the gap between what you want and what you have, to connect the unknown with the data, the conclusion with the hypothesis. Could you restate your problem so that unknown and data, hypothesis and conclusion appear *nearer to each other?*

Transform the conclusion, or transform the hypothesis, or transform both, but so that you bring them nearer to each other. Transform the unknown, or transform the data or the condition, or transform the whole problem, but so that you bring the unknown and the data nearer to each other.

As the solution progresses, new lines appear in the figure considered; new materials and new relations are added to the growing structure in the problem solver's mind. Each transformation of the problem is apt to bring in new elements. *Going back to definitions* is an important way of adding new material to the problem's conception.

For instance, our problem deals with the frustum of a pyramid (as in chapter 7). What is the frustum of a pyramid? How is it *defined?* The frustrum is the portion of a full pyramid that remains when a smaller pyramid is cut off from it by a plane parallel to its base. This answer brings two new solids to our attention, the full pyramid and the small pyramid, and we may find it advantageous to incorporate one or the other or both in the problem's conception.[2]

In going back to the definition of the elements given with the problem we introduce new elements, which in turn introduce still more new elements, and continuing thus we may further *unfold* the problem's concept. Such unfolding often brings us closer to the solution, but it need not; sometimes it may encumber the problem with unnecessary detail.

There are many ways of transforming problems which would deserve consideration; some are applicable only to certain particular kinds of problems, but others are rather general. (See ex. 12.1, 12.2.)

12.6. Wanted: relevant knowledge

The solution consists essentially in connecting the proposed problem with appropriate elements of our formerly acquired knowledge. When we are trying to restate the problem in a more promising form, we are in fact seeking a connection in starting from the problem, we are trying to pierce the clouds surrounding the problem by working at it "from inside." We may seek the connection from the other end, in trying to find some usable piece of knowledge in working at the problem "from outside."

It is obviously impossible to survey all our formerly acquired knowledge. Therefore, we should start exploring the parts of our knowledge that are most likely to be relevant to our problem.

If you are familiar with the domain to which your problem belongs, you know its "key facts," the facts you had most opportunity to use.

[2] *Cf.* HSI, Definition, pp. 85–92.

Make them ready as a good workman makes his favorite tools ready for use so that they are within easy reach.

If you have a problem to find, problems with the same unknown deserve your especial interest; such a problem may be the starting point for regressive planning (see chapter 8). If you have a problem to prove, theorems having the same conclusion as the theorem you are concerned with deserve your especial interest as possible starting points.

What are the key facts here? Is there a problem (especially a formerly solved problem) *with the same kind of unknown? Is there a theorem* (especially a formerly proved theorem) *with the same conclusion?* These questions have a good chance to extract some usable element from your previously acquired knowledge; it is advisable to start with them when you wish to collect pertinent facts. If they remain barren, however, you may be obliged to examine more complex or more recondite facts, or previously considered problems which have some other element in common with your problem, not just the unknown or the conclusion. There are, no doubt, elements in your knowledge which you could use for your present problem—but how are they related to it? How should you reach them? You may try generalization, specialization, analogy; you may ransack the whole province of knowledge to which your problem belongs.

Of course, the more extensive your knowledge is and the better it is organized, the more chance you have to find what you need. See ex. 12.3.

12.7. Wanted: reassessment of the situation

You are disappointed with the progress of your work. Various ideas you have had fizzled out, various routes you have tried led to an impasse. The figure in front of you, the whole conception of the problem in its present state, is perplexing and obscure, crowded but still incomplete; some essential element, some essential link is missing.

The trouble may be that you are involved in side issues and burdened with irrelevant material. Try to go back to the bare conception of the problem; look again at the unknown, the data and the condition, or the hypothesis and the conclusion. *Did you take into account the whole condition? Did you use all the data? Did you take into account the whole hypothesis,* did you use every part of it?

These questions are particularly relevant if you have previously convinced yourself that all the data and all the parts of the condition are needed to determine the unknown, or that the whole hypothesis is needed to derive the conclusion. Yet even if you have no such definite knowledge and you just suspect that all the data and all the clauses of the condition, or of the hypothesis, may be essential, the foregoing questions are justified and may be helpful. They suggest that you should *try to use* a datum or clause which you have neglected so far, and so they may lead you to the missing link.

Or the trouble may be that you have not realized sufficiently the meaning of the essential terms of the problem. *Do you understand— have you visualized—all concepts essentially involved in the problem?* This question may prompt you to *go back to the definition* of some terms, and so it may suggest that you should unfold your problem; it has a chance to lead you to a more satisfactory restatement, to usable new elements.

12.8. The art of asking questions

In the foregoing sections we surveyed typical mental operations or "moves" of the problem solver. The description of each move culminated in a question (or advice—in *italics*) which may serve as an epitome, a condensed expression of the move.[3] It is important to understand how the problem solver (or the teacher) can use these questions.

Each of the collected questions, if asked at the right place and in the right time, may evoke the right answer, the right idea, a well-adapted move which can advance the solution. Thus the question may act as a *stimulant,* prompting the desirable reaction. These questions are *idea-needlers.*

Of course, under given circumstances you may not know what question to ask. Yet you may try several questions, one after the other, and arrive at a question that helps. Thus you can use the foregoing sections as a *repertory* of eligible questions, as a *checklist.*

Do not, however, use this checklist in a haphazard way, taking the questions at random, and do not use it mechanically, going through the questions in a fixed order. Instead, use this list of questions as an expert workman uses his *tool chest.* He takes a good look at the work he has to do and then he selects his tools. He may be obliged to try several tools until he finds the right one, but he does not pick his tools at random, or in a mechanically fixed order; he selects his tools *with judgement.* That is the way how you should select a question from those collected in this chapter when you are facing a problem.

Of course the workman probably has acquired his workmanship by long experience and by watching other workmen carefully. That is

[3] The questions concerned with "problems to find" are listed HSI pp. XVI–XVII (immediately inside the cover of the hard-cover edition). The reader is urged to study this list and the pertaining explanations and illustrations.

how you may master the use of the questions collected here. There is no hard and fast rule directing their use. Yet, if your personal experience of success and failure is behind these questions, and if you are keenly aware of your aim, there is a good chance that you will pick a good one.

Of course, the tools of a good workman are in good repair and well arranged in his tool chest. If you know the questions and the underlying moves described in this chapter, not from description but from experience, and if you have well understood their respective roles in the problem solving process, there is a good chance, I think, that you will handle your problems more professionally, less clumsily, and in a less haphazard way than most people.

Perhaps, any kind of mental discipline consists in the possession and the proper use of a set of appropriate questions. Yet how can we learn the art of asking questions? Has this art no rules?

Examples and Comments on Chapter 12

12.1. *Restate the problem.* The aim of our problem is to prove (or disprove) the statement: "If A is true then B is true." It may be advantageous to transform the problem and try to prove (or disprove) the equivalent contrapositive statement: "If B is false then A is false." See ex. 9.10.

Here is an analogous situation. Let x denote the unknown of a problem to find and $a, b, c, \ldots, l$ the data. (For instance, the unknown and the data may be measurements of various parts of a geometric figure.) It may be advantageous to interchange the unknown x and one of the data, say a. By so doing we pass from the original problem to a new one of which the unknown is a and the data $x, b, c, \ldots, l$. See ex. 2.33, 2.34, and 2.35.

We have considered here two types of transformations *independent of the subject matter.* The study of such types belongs properly to heuristics.

12.2. *Express it in mathematical language.* Descartes's great project which we discussed in sect. 2.1 may be (roughly) condensed into one piece of advice: Transform your problem, whatever it may be, into a mathematical problem by expressing it in the form of algebraic equations. Descartes's project failed, but we may revive it by generalizing it: *Express your problem in mathematical language.* The success of this advice depends, of course, on the scope of the mathematical language at our disposal. For instance, if we know and can use not only the symbols of algebra as Descartes, but also the symbols of the differential and integral calculus, we can treat many more problems.

"Mathematical language," if taken in a very wide sense, may include every kind of sufficiently clarified concept formation. In this very wide interpretation the advice "Express it in mathematical language" may be theoretically perfect, but it is practically pointless: it means no more than "try to be clear."

There is, however, a much narrower and even somewhat vague interpretation which is often useful. Graphs, diagrams, or geometric figures used as symbols form a sort of mathematical language. To *draw a figure*, to express the problem in the language of geometric figures, is often helpful. Some people have an urge to represent their ideas by some sort of geometric symbols. *Cf.* ex. 14.8.

12.3. *A well-stocked and well-organized store* of knowledge is an important asset of the problem solver. Good organization which renders the knowledge readily available may be even more important than the extent of knowledge. Anyhow, there are cases where too much knowledge may be an obstacle; it may prevent the problem solver from seeing a simple approach. Yet good organization can only help.

In a well-arranged store the items most frequently demanded are in the most accessible places, the items often used together are stored together, and labeling and details of arrangement are so planned that we can conveniently assemble any two, or more, related items.

A sensible arrangement of the books in your library or the tools in your tool chest may be helpful, but a sensible arrangement of the knowledge in your memory may be still more helpful and may deserve more care. Let us look at some points of organization important for the problem solver.

(1) In any subject matter there are some *key facts* (key problems, key theorems) which should be stored somehow in the forefront of your memory. When you are starting a problem, you should have some key facts around you, close at hand, just as an expert workman lays out his most frequently used tools around him when he starts working.

When you intend to prove a proposition of elementary plane geometry in the manner of Euclid, you may regard as key facts the four cases of congruence and the four cases of similarity of triangles. When you intend to reduce a problem of elementary geometry to equations in the Cartesian manner (chapter 2), you may regard as key facts the theorem of Pythagoras and the proportionality of the sides in similar triangles. You name the relevant key facts if you are experienced in dealing with the convergence of series, or with some other class of problems.

(2) Two questions which are again and again useful to the problem solver are: *By what data can you determine this kind of unknown? From what hypothesis can you derive this conclusion?* In view of the continual use of these questions, formerly solved problems with the same kind of unknown should be somehow "stored together," as should formerly proved theorems with the same conclusion.

(3) Do you know the city you live in? If you know it very well, you should be able to find the shortest route and the most convenient means of transportation between any two points of the city. Such is the desirable organization of knowledge; in the domain in which you work you should be able to find a practicable connection between any two points.

Surveying related problems may make for a better organization. Thus, the first part of this book has extensively surveyed problems related to each other by the common pattern of their solution. We may also survey strings of problems related by a common unknown, or by common data, or by analogy, and so on.

(4) Euclid wrote not only the "Elements" but also several other works. One

of them, the "Data," surveys the various data by which geometric objects can be determined. I like to believe that Euclid wrote the "Data" to help the problem solver by storing geometric knowledge in a *form readily available* to readers who often ask themselves: *by what data could I determine this kind of unknown?*

12.4. *By what data can you determine this kind of unknown?* List simple "problems to find," whose unknowns are described by one of the following sentences. (Capitals denote points.)

(1) ... find the point *P*.
(2) ... find the length *AB*.
(3) ... find the area of $\triangle ABC$.
(4) ... find the volume of the tetrahedron *ABCD*.

12.5. *From what hypothesis can you derive this conclusion?* List simple theorems of plane geometry whose conclusions coincide with one of the following. (Capitals *A*, *B*, *C*, ... denote points.)

(1) ... then $AB = EF$.
(2) ... then $\angle ABC = \angle EFG$.
(3) ... then $AB:CD = EF:GH$.
(4) ... then $AB < AC$.

12.6. *Analogy: The triangle and the tetrahedron.* Here is a pair of problems, the first about the triangle, the second about the tetrahedron, which are analogous to each other:

Inscribe a circle in a given triangle.
Inscribe a sphere in a given tetrahedron.

List more pairs of problems, or pairs of theorems, similarly related to each other. Are the solutions, or proofs, also analogous, or how are they related to each other?

12.7. State a theorem about triangles analogous to the following theorem about tetrahedra:

A line joining the midpoints of two opposite edges of a tetrahedron passes through the center of gravity of any cross-section of the tetrahedron that is parallel to those two edges.

Could the theorem on the triangle help to prove the theorem on the tetrahedron? Also answer the corresponding questions dealing with the theorems in ex. 12.8 and ex. 12.9.

12.8. (Continued) Any plane passing through the midpoints of two opposite edges of a tetrahedron bisects its volume.

12.9. (Continued) In any tetrahedron the plane bisecting a dihedral angle divides the opposite edge into segments that are proportional to the areas of the adjacent faces including the dihedral angle.

12.10. *Attention and action.*

(1) Does method consists entirely in paying attention to all relevant points, one after the other, in appropriate sequence? (See the motto of this chapter.) I would not dare to assert that. Yet, certainly, a good part of the methodical work of the problem solver consists in focusing the relevant elements of his problem and their various combinations, one after the other.

The question, *"What is the unknown?"* and the advice, *"Look at the unknown!"* aim at the same effect: to turn the problem solver's attention to the unknown of his problem. Working methodically, the problem solver proceeds so as if he were directed by an inner voice:

Look at the problem as a whole.
Look at the unknown.
Look at the data.
Look at the condition.
Look at each datum separately.
Look at each clause of the condition separately.
Look especially at the datum that you have not used yet.
Look especially at the clause of the condition that you have not used yet.
Look at the combination of these two data.
And so on.

(2) Attention may initiate action.

Look at the unknown! What is the unknown? How can you find this kind of unknown? By what data can you determine this kind of unknown? Do you know— have you solved—a problem with this kind of unknown? The attention given the unknown induces the problem solver to search his memory for formerly solved problems with the same unknown. If the search succeeds, the problem solver may attempt to solve his problem in working backward (see chapter 8).

The case considered (recursive work initiated by attention paid to the unknown) is particularly frequent and useful, but attention given to any relevant element of the problem may lead to a profitable contact and hence to profitable action. For instance, attention to a term arising in the formulation of the problem may lead to *going back to the definition* of that term and so to a helpful restatement of the problem, to the introduction of usable new elements into the restated problem.

(3) The problem solver pays attention successively to various elements of the problem and to various combinations of these elements—he hopes to detect one that gives access to some profitable action—or to detect the one that gives access to the most profitable action. He hopes to find a bright idea that shows in a flash what there is to do.

12.11. *Productive thinking, creative thinking.* Thinking is productive if it produces the solution of the present problem, thinking is creative if it creates means to solve future problems. The greater the number and variety of the problems to which the means created are applicable, the greater is the creativity.

The problem solver may do creative work even if he does not succeed in solving his own problem; his effort may lead him to means applicable to other problems.

Then the problem solver may be creative *indirectly,* by leaving a good unsolved problem which eventually leads others to discovering fertile means.

I think that the Greeks who left us the problem of the trisection of an arbitrary angle did great creative work although they did not solve this problem, and inspite of the fact that, through the intervening centuries, this problem occasioned a tremendous amount of unproductive work.   Yet this problem revealed a contrast: whereas an arbitrary angle can be bisected, only some special angles (such as 90°) can be easily trisected by ruler and compasses; this led to the problem of dividing an angle into 5, 7, or 17 equal parts, contributed to the problem of solving equations by radicals, and eventually to the discoveries of Gauss, Abel, and Galois—which created means applicable to countless problems unsuspected by the Greek who first meditated on the trisection of an angle.

CHAPTER 13

RULES
OF DISCOVERY ?

*Though it is difficult to prescribe any Thing in these Sorts of Cases,
and every Person's own Genius ought to be his Guide in these
Operations; yet I will endeavour to show the Way to Learners.*
NEWTON: *Universal Arithmetick* translated by Ralphson, 1769, p. 198.

*Since Arts are more easily learnt by Examples than Precepts, I have
thought fit to adjoin the Solution of the following Problems.*
NEWTON: *op. cit.,* pp. 177–178.

13.1. Rules and rules

As the work of the problem solver progresses, the face of the problem
continually changes. At each stage, the problem solver is confronted
with a new situation and a new decision: what should he do in this
situation, what should be the next step? *If* he possessed a perfect
method, an infallible strategy of problem solving, he could determine
the next step from the data of the present situation by clear reasoning,
on the basis of precise rules. Unfortunately there is no universal perfect method of problem solving, there are no precise rules applicable
to all situations, and in all probability there will never be such rules.

Yet there are rules and rules. Rules of conduct, maxims, and guidelines can be quite useful and reasonable without being as strict as the
rules of mathematics or logic. A mathematical rule is like a mathematical "line without breadth" separating black from white. Yet there
are quite reasonable rules which leave some latitude, some room for
secondary considerations; there is no sharp line of demarcation, and
sometimes there is no black and no white, just various shades of grey.

89

There seem to be attitudes, modes of thinking, habits of mind useful in many, and perhaps in most, problem solving situations. At any rate, the examples and discussions of the foregoing chapters seem to suggest such attitudes. Therefore, we should not ask: "Are there rules of discovery?" but we should put the question differently, perhaps so: "Are there maxims of some sort expressing attitudes useful in problem solving?"[1]

13.2. Rationality

We call an act or a belief *rational* if it is based on well-considered clear reasons and does not spring merely from less articulate and more obscure sources such as habit, unscrutinized impressions, feelings, or "inspiration." The assent we accord to a mathematical theorem after having critically examined its demonstration step by step, is the prototype of rational belief. From a certain point of view the main merit of the study of mathematical demonstrations is that it brings us nearer to that ideal rational behavior that befits man, the "rational being."

It is not obvious, however, how the problem solver should behave rationally. Let us consider his difficulty a little more concretely, let us visualize a frequently arising typical situation. It occurs to a problem solver as he is working at a certain problem A, that this problem is connected with another problem B. The study of problem B could possibly bring him nearer to his goal, the solution of his original problem A. On the other hand, however, the study of problem B may remain sterile and could result in loss of time and effort. And so the problem solver faces a decision: should he abandon for a time the study of his original problem A and switch to the study of the new problem B? His dilemma is to introduce or not to introduce B as an auxiliary problem. Can the problem solver arrive at a rational decision?

One important benefit that the problem solver may derive from problem B is that work on B may stir his memory and bring to the surface elements that he can use in solving his original problem A. What are the chances that work on B will have this desirable effect? It seems impossible to evaluate such chances merely on the basis of some clear, rational argument; to some extent, in some manner the problem solver must rely on his inarticulate feelings.

On the other hand, there may be articulate reasons for or against introducing B as an auxiliary problem; we have surveyed some such reasons in chapter 9. How should the problem solver take into account both his inarticulate feelings and his articulate reasons? He may—and

---

[1] The majority of the "rules" discussed in this chapter have been introduced, in a somewhat different formulation, in paper no. 13 of the author; see the Bibliography.

this seems to be a sensible procedure—carefully consider his articulate reasons for an appropriate time and refer to his feelings afterwards for the final decision. In fact, the careful consideration of his articulate reasons has had a chance to influence his feelings in the right direction; he could scarcely have acted more rationally.

At any rate, the problem solver must learn to balance inarticulate feelings against articulate reasons. And perhaps this is the most important thing he must learn. It seems to me that the main rule of conduct, the principal maxim of the problem solver should be:

Never act against your feelings, but try to see with an open mind clear reasons for or against your plans.

13.3. Economy, but no predeterminable limitation

Tendency to economy needs no explanation. Everybody understands that you wish to save your assets, that you try to expend as little money, time, and effort as possible in performing a given task. Your mind may be your most important asset, and saving mental effort may be the most important kind of economy. *Do not do with more what can be done with less.* This is the general principle of economy; we observe it in problem solving if we try to attain the solution by bringing in *as little extraneous material as possible.*

Obviously, we have to scrutinize carefully the problem itself and such materials as immediately belong to it; we should first try to see a path to the solution without examining other things. If we fail to notice such a path, we examine materials which do not belong so immediately to the problem but are still close to it. If we fail again to notice some useful suggestion we may proceed to further details but—and this is the desirable general attitude—we are reluctant to spend time and effort on farfetched things as long as there is some hope that we can solve the proposed problem by starting from things closer to it. This sort of common sense *economy* should be expressed by the maxim:

Stay as close to the problem as possible.

Yet we cannot predict how close to the problem we shall be able to stay. Some superior being, possessing a perfect method of problem solving unattainable to us, could foretell with certainty how far away he will be obliged to go to collect the materials necessary to the solution —yet we cannot. In accordance with the principle of economy, we first explore the proposed problem itself; if this is not enough, we explore the immediate neighborhood of the problem. If even this is not enough, we explore a wider neighborhood; whenever our exploration fails to discover a path to the solution, we are obliged to go further. If you have made up your mind to solve the problem at any cost, you may pay a very high cost indeed; at any rate, you cannot limit the cost in advance. The resolute problem solver has to accept the principle of *absence of limitation,* counterpart of the principle of economy:

Yet be prepared to go as far away from the problem as circumstances may oblige you to go.

13.4. Persistence, but variety

"Genius is patience."

"Genius is one per cent inspiration and ninety-nine per cent perspiration."

One of these sayings is attributed to Buffon, the other to Edison, and both convey the same message: a good problem solver must be obstinate, he must stick to his problem, he must not give up.

What applies to the whole does not quite apply to the parts. Certainly, the problem solver examining some detail or some aspect of his problem should stick to it, should not give up too soon. Yet he should also try to assess the prospects and should not insist on squeezing an orange he has already squeezed completely dry.

Stick to the point examined till there is hope for some useful suggestion.

The work of the problem solver is largely a work of mobilization; he has to extract from his memory items applicable to his present problem. The item he needs to recall may be closer associated with, and more easily remembered through, a certain aspect or a certain detail of his problem than through other aspects or details. But the problem solver does not know in advance which detail or aspect will bring him nearer to his goal. Therefore, he should consider a variety of aspects or details, and he should certainly consider all the more basic or more promising ones.

To cover the whole territory without loss of time, the problem solver should not stay too long in, or return too soon to, the same spot. He should seek variety, he should see something new at each stage—a new point, or a new combination of points previously examined, or see in a new light points and combinations already considered. The aim is, of course, to see the whole problem in a new, more promising light.

In brief, variety is a necessary counterpart of persistence. As stated earlier, you should explore the points you consider with some persistence. *Yet try to examine some ground not yet covered at each step and try to perceive some useful suggestion in whatever you examine.*

The most obvious danger that this maxim warns you of, the greatest enemy of variety, is falling in a groove—repeating the same thing over and over again without change and without progress.

13.5. Rules of preference

If there are two approaches to the same problem which appear equally advantageous in other respects, but one of which seems easier than the other, it is natural to try the easier approach first. We see here a (rather trivial) *rule of preference* which we may state:

The less difficult precedes the more difficult.

In fact, this statement as it stands is incomplete: we should have added as a limitation or restriction *"ceteris paribus"* or "other things being equal." Now, let us note that this essential limitation, although not expressed, must be *understood* in this and in all subsequent similarly formulated rules of preference. Following are a couple equally obvious rules of this kind:

The more familiar precedes the less familiar.

An item having more points in common with the problem precedes an item having less such points.

These rules are obvious, but their application may be less obvious. The restriction *"ceteris paribus,"* especially, not expressed but understood, may require all the subtlety of the problem solver.

There are other less obvious, less general, more specific rules of preference. To examine them in an orderly fashion we should classify the elements involved. Here is a classification which may be incomplete, but at least the most conspicuous cases fit quite naturally into it:

(1) Materials inherent in the problem.
(2) Available knowledge.
(3) Auxiliary problems.

We shall take up the connected rules in the three following sections.

13.6. Materials inherent in the problem

When you start examining a problem, you do not know yet which details of the problem will turn out important. Hence there is the danger that you may emphasize too much an unimportant detail, and then you may get caught by it, you cannot get away from it. Therefore, begin by examining the problem as an undivided whole, do not concern yourself with details, let the whole of the problem work on your mind until you fully realize the point, the aim of the problem. *The whole precedes the parts.*

When you have the impression that you cannot profit much more by the consideration of the problem as a whole and you are about to go into details, observe that there is something like a hierarchy of details.

The highest ranking, nearest to the "center" of the problem, are the principal parts. (As we have discussed, the hypothesis and the conclusion are the principal parts of a problem to prove, the unknown, the data, and the condition those of a problem to find.) It is natural to begin the detailed consideration of the problem with the principal parts; you should see clearly, very clearly, the desired conclusion and the hypothesis from which it should follow—or the desired unknown, the available data, and the connecting condition. *The principal parts precede the other parts.*

One or the other of the principal parts may be subdivided: the hypothesis may consist of several assumptions, the conclusion of several assertions, and the condition of several clauses; the unknown may be a multipartite unknown having several components; and there may be several data which you may have lumped together in a first consideration. After the principal parts, their subdivisions deserve your next attention; you may examine each of the data, each of the unknowns, each clause of the condition, each assumption of the hypothesis, each assertion of the conclusion by itself. At any rate, other details of the problem may be considered as further removed from its center than the principal parts, which rank highest, and the subdivisions of the principal parts, which come next. There may be an order of precedence also among remoter parts of the problem. (Thus, a certain concept A may be involved in the statement of the problem, and another concept B may be involved in the definition of A; obviously B is further removed from the center of the problem than A.) Now, do not go further than necessary. Other things being equal (this limitation remains with us), you have more chance to make good use of a part closer to the center of the problem than of a more remote one. *Less remote parts precede more remote parts.*

13.7. Available knowledge

As we have repeatedly discussed, an essential (perhaps the most essential) performance of the problem solver is to mobilize the relevant elements of his knowledge and connect them with the elements of his problem. The problem solver can work at this task "from inside" or "from outside." He may remain inside his problem, unfolding it, scrutinizing its various parts, and hoping that such scrutiny will attract some usable piece of knowledge. Or he may go outside his problem, roaming over various regions of his knowledge and looking for usable pieces. In the foregoing section we have observed the problem solver working from inside, now we wish to see him working from outside.

Any piece of knowledge, any experience of our past could be usable for the solution of the problem at hand, but it is obviously impossible to pass in review all our knowledge or to recall our whole past point by point. Even if our problem is mathematical and we take into account only that relatively clear and well-ordered portion of our knowledge that consists of formerly solved problems and formerly proved theorems within a certain branch of mathematics, we cannot undertake to examine all that material item by item. We have to restrict ourselves, we have to pick out such items as have the best chance to be usable.

Let us consider, one after the other, problems to solve and problems to prove.

We have a problem to solve. We have already considered its principal parts, the unknown, the data, the condition. Now we are searching our memory for some formerly solved problem that could be helpful. It is natural to look for one that has something in common with our present problem, the unknown, or one of the unknowns, the data, or one of the data, some concept essentially involved, and so on. There is a possibility, more or less remote, that any such formerly solved problem may be helpful, but there are certainly too many for us to examine. Yet, among all the possible points of contact there is one that deserves more attention than the others—the *unknown*. (Especially, we may try to use a formerly solved problem that has the same kind of unknown as our present problem as the starting point of a regressive solution, of working backward; see chapter 8.) Of course, in special situations other points of contact may be preferable, but in general, *a priori,* other things being equal, we should first *look at the unknown.*[2] *Formerly solved problems having the same kind of unknown as the present problem precede other formerly solved problems.*

If we cannot find a sufficiently approachable formerly solved problem that has the same kind of unknown as the problem in hand we may look for one that has a *similar* kind of unknown—even such problems have a high priority, although not the highest priority.

If we have a problem to prove, the situation is similar. In searching our memory for a formerly proved theorem that could be helpful, we should *look at the conclusion. Formerly proved theorems with the same conclusion as the theorem that we are trying to prove precede other formerly proved theorems.*

The next best thing to a formerly proved theorem with the same conclusion may be one with a conclusion similar to the conclusion of the theorem we want to establish.

[2] HSI, pp. 123–129, Look at the unknown.

13.8. Auxiliary problems

One of the most crucial decisions confronting the problem solver is the choice of an appropriate auxiliary problem. He may look for such a problem working from inside, or working from outside, or (what is often the most sensible procedure) working alternately from inside and outside. Certain sorts of auxiliary problems, other things being equal, have more chance to be useful than others.

An auxiliary problem may advance the solution of the proposed problem in an inexhaustible variety of ways: it may yield material help, methodological help, stimulating influence, guidance, or practice. Yet whatever kind of help we are seeking, we have more chance *a priori* to obtain it from an auxiliary problem whose connection with the proposed problem is closer, than from another whose connection is looser. *Problems equivalent to the proposed one precede such problems as are more or less ambitious, and these precede the rest.* We can express the same thing in other terms: *Bilateral reduction precedes unilateral reduction which precedes looser connections* (see chapter 9).

13.9. Summary

Rationality. *Never act against your feelings, but try to see with an open mind clear reasons for or against your plans.*

Economy but no predeterminable limitation. *Stay as close to the problem as possible. Yet be prepared to go as far away from it as circumstances may oblige you to go.*

Persistence but variety. *Stick to the point examined until there is hope for some useful suggestion. Yet try to examine some ground not yet covered at each step and try to perceive some useful suggestion in whatever you examine.*

Rules of preference

The less difficult precedes the more difficult.

The more familiar precedes the less familiar.

An item having more points in common with the problem precedes an item having less such points.

The whole precedes the parts, the principal parts precede the other parts, less remote parts precede more remote parts.

Formerly solved problems having the same kind of unknown as the present problem precede other formerly solved problems.

Formerly proved theorems with the same conclusion as the theorem that we are trying to prove precede other formerly proved theorems.

Problems equivalent to the proposed one precede such problems as are more or less ambitious, and these precede the rest. Or: *Bilateral reduction precedes unilateral reduction which precedes looser connections.*

Yet add mentally to all rules of preference: OTHER THINGS BEING EQUAL.

Examples and Comments on Chapter 13

13.1. *The genius, the expert, and the beginner.* The genius acts according to the rules without knowing that there are rules. The expert acts according to the rules without thinking of the rules, but, if needed, he could quote the rule applicable to the case in question. The beginner, trying to act according to the rule, may learn its veritable meaning from success and failure.

These remarks are not new of course. St. Augustine, speaking of orators and the rules of rhetoric, put it so: "They follow the rules because they are eloquent, they are not eloquent because they follow the rules."

13.2. *Of plums and plans.* Shall I take this fruit? Is it ripe enough to be picked? Of course, if I leave it on the tree, it may still ripen and gain in flavor. On the other hand, if it remains on the tree, it may be attacked by birds or insects, knocked down by the wind, taken by the neighborhood children, or be spoiled or lost in some other way. Should I take it now? Has it quite the right color, shape, softness, smell, general appearance?

Color, shape, softness, and smell indicate the flavor, they do not guarantee it. When examining a fruit from my own garden, I can assess such indications quite reliably, at least I think so. When the fruit is not so familiar to me, my estimate is certainly much less reliable. Anyhow, estimating the taste of a fruit on the basis of its appearance can scarcely be considered as "fully objective." Such estimates depend to a large extent on personal experience which can scarcely be explained completely or argued exhaustively on an impersonal level.

Shall I take this step? Is my plan ripe enough to be carried through? Of course it is not certain that my plan will work. If I pondered it a little longer, I could see its chances a little better. On the other hand, I should do something sooner or later and, for the moment, I cannot think of a more reliable plan. Should I start carrying through that plan now? Does it look promising enough?

In picking plums or pursuing plans we may have some articulate reasons, but our decision will scarcely be based on reason alone. Our evaluation of the general aspect, of the promise of the fruit or of the problem situation, is bound to depend on feelings not fully analyzable.

13.3. *Style of work.* Anybody who intends to formulate "rules of discovery" should realize that different problem solvers work differently. Each good problem solver has his individual style.

Let us compare two problem solvers; one has the mentality of an "engineer," the other the mentality of a "physicist." They are trying to solve the same problem, but they are working differently as they have different interests. The engineer aims at a neat, short, efficient solution (the "least expensive," the "most marketable" solution). The physicist aims at the principle underlying the solution. The engineer is more inclined to "productive," the physicist to "creative," thinking (ex. 12.11). Accordingly they prefer different means to the same end.

Let us consider a half-concrete example. There seem to be two approaches to the problem that the engineer and the physicist are trying to solve. On the one hand, the proposed problem shows some analogy to a certain formerly solved problem A. On the other hand, the proposed problem seems to be amenable to treatment according to a general pattern B. There is a choice between these two approaches, between A and B. I am inclined to think that under such circumstances, other things being equal, the engineer prefers to investigate the concrete problem A and the physicist the general pattern B.

This example leads to the general statement: the style of work of the problem solver consists essentially in a *system of preferences,* of priorities. To the rules of preference summarized in sect. 13.9, the problem solver may add others (for instance: "General patterns precede particular facts"). Moreover, he may *emphasize* certain rules of preference more than others ("When there is a conflict, rule X has more weight than rule Y").

CHAPTER 14

ON LEARNING, TEACHING, AND LEARNING TEACHING[1]

What you have been obliged to discover by yourself leaves a path in your mind which you can use again when the need arises.

G. C. LICHTENBERG: *Aphorismen.*

Thus all human cognition begins with intuitions, proceeds from thence to conceptions, and ends with ideas.

I. KANT: *Critique of Pure Reason,* translated by
J. M. D. Meiklejohn, 1878, p. 429.

I [planned to] write so that the learner may always see the inner ground of the things he learns, even so that the source of the invention may appear, and therefore in such a way that the learner may understand everything as if he had invented it by himself.

G. W. VON LEIBNITZ: *Mathematische Schriften,* edited by Gerhardt,
vol. VII, p. 9.

14.1. Teaching is not a science

I shall tell you some of my opinions on the process of learning, on the art of teaching, and on teacher training.

My opinions are the result of a long experience. Still, such personal opinions may be irrelevant and I would not dare to waste your time by telling them if teaching could be fully regulated by scientific facts and

---

[1] Sect. 14.1–14.7 were presented as an address at the 46th Annual Meeting of the Mathematical Association of America at Berkeley and printed previously; see the Bibliography, paper no. 20 of the author.

theories. This, however, is not the case. Teaching is, in my opinion, not just a branch of applied psychology—at any rate, it is not yet that for the present.

Teaching is correlated with learning. The experimental and theoretical study of learning is an extensively and intensively cultivated branch of psychology. Yet there is a difference. We are principally concerned here with complex learning situations, such as learning algebra or learning teaching, and their long-term educational effects. The psychologists, however, devote most of their attention to, and do their best work about, simplified short-term situations. Thus the psychology of learning may give us interesting hints, but it cannot pretend to pass ultimate judgment upon problems of teaching.[2]

14.2. The aim of teaching

We cannot judge the teacher's performance if we do not know the teacher's aim. We cannot meaningfully discuss teaching, if we do not agree to some extent about the aim of teaching.

Let me be specific. I am concerned here with mathematics in the high school curriculum and I have an old fashioned idea about its aim: first and foremost, it should teach those young people to THINK.

This is my firm conviction; you may not go along with it all the way, but I assume that you agree with it to some extent. If you do not regard "teaching to think" as a primary aim, you may regard it as a secondary aim—and then we have enough common ground for the following discussion.

"Teaching to think" means that the mathematics teacher should not merely impart information, but should try also to develop the ability of the students to use the information imparted: he should stress know-how, useful attitudes, desirable habits of mind. This aim may need fuller explanation (my whole printed work on teaching may be regarded as a fuller explanation) but here it will be enough to emphasize only two points.

First, the thinking with which we are concerned here is not day-dreaming but "thinking for a purpose" or "voluntary thinking" (William James) or "productive thinking" (Max Wertheimer). Such "thinking" may be identified here, at least in first approximation, with "problem solving." At any rate, in my opinion, one of the principal aims of the high school mathematics curriculum is to develop the students' ability to solve problems.

Second, mathematical thinking is not purely "formal"; it is not con-

---

[2] E. R. Hilgard, *Theories of Learning*, 2nd ed., Appleton-Century-Crofts, New York, 1956. *Cf.* pp. 485–490.

cerned only with axioms, definitions, and strict proofs, but many other things belong to it: generalizing from observed cases, inductive arguments, arguments from analogy, recognizing a mathematical concept in, or extracting it from, a concrete situation. The mathematics teacher has an excellent opportunity to acquaint his students with these highly important "informal" thought processes, and I mean that he should use this opportunity better, and much better, than he does today. Stated incompletely but concisely: let us teach proving by all means, but let us also teach guessing.

14.3. Teaching is an art

Teaching is not a science, but an art. This opinion has been expressed by so many people so many times that I feel a little embarrassed repeating it. If, however, we leave a somewhat hackneyed generality and get down to appropriate particulars, we may see a few tricks of our trade in an instructive sidelight.

Teaching obviously has much in common with the theatrical art. For instance, you have to present to your class a proof which you know thoroughly having presented it already so many times in former years in the same course. You really cannot be excited about this proof—but, please, do not show that to your class; if you appear bored, the whole class will be bored. Pretend to be excited about the proof when you start it, pretend to have bright ideas when you proceed, pretend to be surprised and elated when the proof ends. You should do a little acting for the sake of your students who may learn, occasionally, more from your attitudes than from the subject matter presented.

I must confess that I take pleasure in a little acting, especially now that I am old and very seldom find something new in mathematics; I may find a little satisfaction in re-enacting how I discovered this or that little point in the past.

Less obviously, teaching has something in common also with music. You know, of course, that the teacher should not say things just once or twice, but three or four or more times. Yet, repeating the same sentence several times without pause and change may be terribly boring and defeat its own purpose. Well, you can learn from the composers how to do it better. One of the principal art forms of music is "air with variations." Transposing this art form from music into teaching you begin by saying your sentence in its simplest form; then you repeat it with a little change; then you repeat it again with a little more color, and so on; you may wind up by returning to the original simple formulation. Another musical art form is the "rondo." Transposing the rondo from music into teaching, you repeat the same essential sentence several times with little or no change, but you insert between two repetitions some appropriately contrasting illustrative material. I hope that when you listen the next time to a theme with variations by Beethoven or to a rondo by Mozart you will give a little thought to improving your teaching.

Now and then, teaching may approach poetry, and now and then it may approach profanity. May I tell you a little story about the great Einstein? I listened once to Einstein as he talked to a group of physicists in a party. "Why have all the electrons the same charge?" said he. "Well, why are all the little balls in the goat dung of the same size?" Why did Einstein say such things? Just to make some snobs to raise their eyebrows? He was not disinclined to do so, I think. Yet, probably, it went deeper. I do not think that the overheard remark of Einstein was quite casual. At any rate, I learnt something from it: Abstractions are important; use all means to make them more tangible. Nothing is too good or too bad, too poetical or too trivial to clarify your abstractions. As Montaigne put it: The truth is such a great thing that we should not disdain any means that could lead to it. Therefore, if the spirit moves you to be a little poetical, or a little profane, in your class, do not have the wrong kind of inhibition.

14.4. Three principles of learning

Teaching is a trade that has innumerable little tricks. Each good teacher has his pet devices and each good teacher is different from any other good teacher.

Any efficient teaching device must be correlated somehow with the nature of the learning process. We do not know too much about the learning process, but even a rough outline of some of its more obvious features may shed some welcome light upon the tricks of our trade. Let me state such a rough outline in the form of three "principles" of learning. Their formulation and combination is of my choice, but the "principles" themselves are by no means new; they have been stated and restated in various forms, they are derived from the experience of the ages, endorsed by the judgment of great men, and also suggested by the psychological study of learning.

These "principles of learning" can be also taken for "principles of teaching," and this is the chief reason for considering them here—but more about this later.

(1) *Active learning.* It has been said by many people in many ways that learning should be active, not merely passive or receptive; merely by reading books or listening to lectures or looking at moving pictures without adding some action of your own mind you can hardly learn anything and certainly you can not learn much.

There is another often expressed (and closely related) opinion: *The best way to learn anything is to discover it by yourself.* Lichtenberg (an eighteenth century German physicist, better known as a writer of aphorisms) adds an interesting point: *What you have been obliged to discover by yourself leaves a path in your mind which you can use again when the need arises.* Less colorful but perhaps more widely applicable, is the following statement: *For efficient learning, the learner should discover by himself as large a fraction of the material to be learned as feasible under the given circumstances.*

This is the *principle of active learning* (Arbeitsprinzip). It is a very old principle: it underlies the idea of "Socratic method."

(2) *Best motivation.* Learning should be active, we have said. Yet the learner will not act if he has no motive to act. He must be induced to act by some stimulus, by the hope of some reward, for instance. The interest of the material to be learned should be the best stimulus to learning and the pleasure of intensive mental activity should be the best reward for such activity. Yet, where we cannot obtain the best we should try to get the second best, or the third best, and less intrinsic motives of learning should not be forgotten.

For efficient learning, the learner should be interested in the material to be learned and find pleasure in the activity of learning. Yet, beside these best motives for learning, there are other motives too, some of them desirable. (Punishment for not learning may be the least desirable motive.)

Let us call this statement the *principle of best motivation.*

(3) *Consecutive phases.* Let us start from an often quoted sentence of Kant: *Thus all human cognition begins with intuitions, proceeds from thence to conceptions, and ends with ideas.* The English translation uses the terms "cognition, intuition, idea." I am not able (who is able?) to tell in what exact sense Kant intended to use these terms. Yet I beg your permission to present my reading of Kant's dictum:

Learning begins with action and perception, proceeds from thence to words and concepts, and should end in desirable mental habits.

To begin with, please, take the terms of this sentence in some sense that you can illustrate concretely on the basis of your own experience. (To induce you to think about your personal experience is one of the desired effects.) "Learning" should remind you of a classroom with yourself in it as student or teacher. "Action and perception" should suggest manipulating and seeing concrete things such as pebbles, or apples, or Cuisenaire rods; or ruler and compasses; or instruments in a laboratory; and so on.

Such concrete interpretation of the terms may come more easily and more naturally when we think of some simple elementary material. Yet after a while we may perceive similar phases in the work spent on mastering more complex, more advanced material. Let us distinguish three phases: the phases of *exploration, formalization,* and *assimilation.*

A first *exploratory* phase is closer to action and perception and moves on a more intuitive, more heuristic level.

A second *formalizing* phase ascends to a more conceptual level, introducing terminology, definitions, proofs.

The phase of *assimilation* comes last: there should be an attempt to perceive the "inner ground" of things, the material learned should be mentally digested, absorbed into the system of knowledge, into the whole mental outlook of the learner; this phase paves the way to applications on one hand, to higher generalizations on the other.

Let us summarize: *For efficient learning, an exploratory phase should precede the phase of verbalization and concept formation and, eventually, the material learned should be merged in, and contribute to, the integral mental attitude of the learner.*

This is the *principle of consecutive phases.*

14.5. Three principles of teaching

The teacher should know about the ways of learning. He should avoid inefficient ways and take advantage of the efficient ways of learning. Thus he can make good use of the three principles we have just surveyed, the principle of active learning, the principle of best motivation, and the principle of consecutive phases; these principles of learning are also principles of teaching. There is, however, a condition: to avail himself of such a principle, the teacher should not merely know it from hearsay, but he should understand it intimately on the basis of his own well-considered personal experience.

(1) *Active learning.* What the teacher says in the classroom is not unimportant, but what the students think is a thousand times more important. The ideas should be born in the students' mind and the teacher should act only as midwife.

This is a classical Socratic precept and the form of teaching best adapted to it is the Socratic dialogue. The high school teacher has a definite advantage over the college instructor in that he can use the dialogue form much more extensively. Unfortunately, even in the high school, time is limited and there is a prescribed material to cover so that all business cannot be transacted in dialogue form. Yet the principle is: let the students *discover by themselves as much as feasible* under the given circumstances.

Much more is feasible than is usually done, I am sure. Let me recommend to you here just one little practical trick: let the students *actively contribute to the formulation* of the problem that they have to solve afterwards. If the students have had a share in proposing the problem they will work at it much more actively afterwards.

In fact, in the work of the scientist, formulating the problem may be the better part of a discovery, the solution often needs less insight and originality than the formulation. Thus, letting your students have a share in the formulation, you not only motivate them to work harder, but you teach them a desirable attitude of mind.

(2) *Best motivation.* The teacher should regard himself as a salesman: he wants to sell some mathematics to the youngsters. Now, if the salesman meets with sales resistance and his prospective customers refuse to buy, he should not lay the whole blame on them. Remember, the customer is always right in principle, and sometimes right in practice. The lad who refuses to learn mathematics may be right; he may be neither lazy nor stupid, just more interested in something else—there are so many interesting things in the world around us. It is your duty as a teacher, as a salesman of knowledge, to convince the student that mathematics *is* interesting, that the point just under discussion is interesting, that the problem he is supposed to do deserves his effort.

Therefore, the teacher should pay attention to the choice, the formulation, and a suitable presentation of the problem he proposes. The problem should appear as meaningful and relevant from the student's standpoint; it should be related, if possible, to the everyday experience of the students, and it should be introduced, if possible, by a little joke or a little paradox. Or the problem should start from some very familiar knowledge; it should have, if possible, some point of general interest or eventual practical use. If we wish to stimulate the student to a genuine effort, we must give him some reason to suspect that his task deserves his effort.

The best motivation is the student's interest in his task. Yet there are other motivations which should not be neglected. Let me recommend here just one little practical trick. Before the students do a problem, let them *guess the result,* or a part of the result. The boy who expresses an opinion commits himself; his prestige and self-esteem depend a little on the outcome, he is impatient to know whether his guess will turn out right or not, and so he will be actively interested in his task and in the work of the class—he will not fall asleep or misbehave.

In fact, in the work of the scientist, the guess almost always precedes the proof. Thus, in letting your students guess the result, you not only motivate them to work harder, but you teach them a desirable attitude of mind.

(3) *Consecutive phases.* The trouble with the usual problem material of the high school textbooks is that they contain almost exclusively merely routine examples. A routine example is a short range example; it illustrates, and offers practice in the application of, just one isolated rule. Such routine examples may be useful and even necessary, I do not deny it, but they miss two important phases of learning: the exploratory phase and the phase of assimilation. Both phases seek to connect the problem in hand with the world around us and with other knowledge, the first before, the last after, the formal solution. Yet the routine problem is obviously connected with the rule it illustrates and it is scarcely connected with anything else, so that there is little profit in seeking further connections. In contrast with such routine problems, the high school should present more challenging problems at least now and then, problems with a rich background that deserve further exploration, and problems which can give a foretaste of the scientist's work.

Here is a practical hint: if the problem you want to discuss with your class is suitable, let your students do some preliminary exploration: it may whet their appetite for the formal solution. And reserve some time for a retrospective discussion of the finished solution; it may help in the solution of later problems.

(4) After this much too incomplete discussion, I must stop explaining the three principles of active learning, best motivation, and consecutive phases. I think that these principles can penetrate the details of the teacher's daily work and make him a better teacher. I think too that these principles should also penetrate the planning of the whole curriculum, the planning of each course of the curriculum, and the planning of each chapter of each course.

Yet it is far from me to say that you must accept these principles. These principles proceed from a certain general outlook, from a certain philosophy, and you may have a different philosophy. Now, in teaching as in several other things, it does not matter much what your philosophy is or is not. It matters more whether you have a philosophy or not. And it matters very much whether you try to live up to your philosophy or not. The only principles of teaching which I thoroughly dislike are those to which people pay only lip service.

14.6. Examples

Examples are better than precepts; let me get down to examples—I much prefer examples to general talk. I am here concerned principally with teaching on the high school level, and I shall present you a few examples on that level. I often find satisfaction in treating examples at the high school level, and I can tell you why: I attempt to treat them so that they recall in one respect or the other my own mathematical experience; I am re-enacting my past work on a reduced scale.

(1) *A seventh grade problem.* The fundamental art form of teaching is the Socratic dialogue. In a junior high school class, perhaps in the seventh grade, the teacher may start the dialogue so:

"What is the time at noon in San Francisco?"

'But, teacher, everybody knows that' may say a lively youngster, or even 'But teacher, you are silly: twelve o'clock.'

"And what is the time at noon in Sacramento?"

'Twelve o'clock—of course, not twelve o'clock midnight.'

"And what is the time at noon in New York?"

'Twelve o'clock.'

"But I thought that San Francisco and New York do not have noon at the same time, and you say that both have noon at twelve o'clock!"

'Well, San Francisco has noon at twelve o'clock Western Standard Time and New York at twelve o'clock Eastern Standard Time.'

"And on what kind of standard time is Sacramento, Eastern or Western?"

'Western, of course.'

"Have the people in San Francisco and Sacramento noon at the same moment?"

"You do not know the answer? Well, try to guess it: does noon come sooner to San Francisco, or to Sacramento, or does it arrive exactly at the same instant at both places?"

How do you like my idea of Socrates talking to seventh grade kids? At any rate, you can imagine the rest. By appropriate questions the teacher, imitating Socrates, should extract several points from the students:

(a) We have to distinguish between "astronomical" noon and conventional or "legal" noon.

(b) Definitions for the two noons.

(c) Understanding "standard time": how and why is the globe's surface subdivided into time zones?

(d) Formulation of the problem: "At what o'clock Western Standard Time is the astronomical noon in San Francisco?"

(e) The only specific datum needed to solve the problem is the longitude of San Francisco (in an approximation sufficient for the seventh grade).

The problem is not too easy. I tried it on two classes; in both classes the participants were high school teachers. One class spent about 25 minutes on the solution, the other 35 minutes.

(2) I must say that this little seventh grade problem has various advantages. Its main advantage may be that it emphasizes an essential mental operation which is sadly neglected by the usual problem material of the textbooks: *recognizing the essential mathematical concept in a concrete situation.* To solve the problem, the students must recognize a *proportionality:* the time of the highest position of the sun in a locality on the globe's surface changes *proportionally* to the longitude of the locality.

In fact, in comparison with the many painfully artificial problems of the high school textbooks, our problem is a perfectly natural, a "real" problem. In the serious problems of applied mathematics, the appropriate *formulation* of the problem is always a major task, and often the most important task; our little problem which can be proposed to an average seventh grade class possesses just this feature. Again, the serious problems of applied mathematics may lead to practical action, for instance, to adopting a better manufacturing process; our little problem can explain to seventh graders why the system of 24 time zones, each with a uniform standard time, was adopted. On the whole, I think that this problem, if handled with a little skill by the teacher, could help a future scientist or engineer to discover his vocation, and it could also contribute to the intellectual maturity of those students who will not use mathematics professionally.

Observe also that this problem illustrates several little tricks mentioned in the foregoing: the students actively contribute to the formulation of the problem [*cf.* sect. 14.5(1)]. In fact, the exploratory phase which leads to the formulation of the problem is prominently important [*cf.* sect. 14.5(3)]. Then, the students are invited to guess an essential point of the solution [*cf.* sect. 14.5(2)].

(3) *A tenth grade problem.* Let us consider another example. Let us start from what is probably the most familiar problem of geometric construction: *construct a triangle, being given its three sides.* As analogy is such a fertile source of invention, it is natural to ask: what is the analogous problem in solid geometry? An average student, who has a little knowledge of solid geometry, may be led to formulate the problem: *construct a tetrahedron, being given its six edges.*

It may be mentioned here parenthetically that this problem of the tetrahedron comes as close as it can on the usual high school level to practical problems solved by "mechanical drawing." Engineers and designers use well executed drawings to give precise information about the details of three dimensional figures of machines or structures to be built: we intend to build a tetrahedron with specified edges. We might wish, for example, to carve it out of wood.

This leads to asking that the problem should be solved precisely, by straight-edge and compasses, and to discussing the question: which details of the tetrahedron should be constructed? Eventually, from a well conducted class discussion, the following definitive formulation of the problem may emerge:

Of the tetrahedron ABCD, we are given the lengths of its six edges

$$AB, BC, CA, AD, BD, CD.$$

Regard $\triangle ABC$ as the base of the tetrahedron and construct with ruler and compasses the angles that the base includes with the other three faces.

The knowledge of these angles is required for cutting out of wood the desired solid. Yet other elements of the tetrahedron may turn up in the discussion such as

(a) the altitude drawn from the vertex D opposite the base,

(b) the foot F of this altitude in the plane of the base;

(a) and (b) would contribute to the knowledge of the solid, they may possibly help to find the required angles, and so we may try to construct them too.

(4) We can, of course, construct the four triangular faces which are assembled in Fig. 14.1. (Short portions of some circles used in the construction are preserved to indicate that $AD_2 = AD_3$, $BD_3 = BD_1$, $CD_1 = CD_2$.) If Fig. 14.1 is copied on cardboard, we can add three flaps, cut out the pattern, fold it along three lines, and paste down the

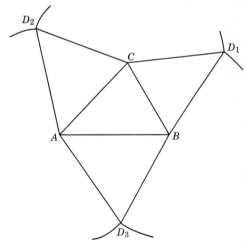

Fig. 14.1. Tetrahedron from six edges.

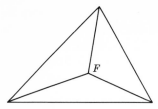

Fig. 14.2. An aspect of the finished product.

flaps; in this way we obtain a solid model on which we can measure roughly the altitude and the angles in question. Such work with cardboard is quite suggestive, but it is not what we are required to do: we should construct the altitude, its foot, and the angles in question with ruler and compasses.

(5) It may help to take the problem, or some part of it, "as solved." Let us visualize how Fig. 14.1 will look when the three lateral faces, after having been rotated each about a side of the base, will be lifted into their proper position. Fig. 14.2 shows the orthogonal projection of the tetrahedron onto the plane of its base, $\triangle ABC$. The point F is the projection of the vertex D: it is the foot of the altitude drawn from D.

(6) We may visualize the transition from Fig. 14.1 to Fig. 14.2 with or without a cardboard model. Let us focus our attention on one of the three lateral faces, on $\triangle BCD_1$, which was originally located in the same plane as $\triangle ABC$, in the plane of Fig. 14.1 which we imagine as horizontal. Let us watch the triangle BCD_1 rotating about its fixed side BC and let our eyes follow its only moving vertex D_1. This vertex D_1 describes an arc of a circle. The center of this circle is a point of BC; the plane of this circle is perpendicular to the horizontal axis of revolution BC; thus D_1 moves in a vertical plane. Therefore, the projection of the path of the moving vertex D_1 onto the horizontal plane of Fig. 14.1 is a straight line, perpendicular to BC, passing through the original position of D_1.

Yet there are two more rotating triangles, three altogether. There are three moving vertices, each following a circular path in a vertical plane—to which destination?

(7) I think that by now the reader has guessed the result (perhaps even before reading the end of the foregoing subsection): the three straight lines drawn from the original positions (see Fig. 14.1) of D_1, D_2, and D_3 perpendicularly to BC, CA, and AB, respectively, meet in one point, the point F, our supplementary aim (b), see Fig. 14.3. (It is enough to draw two perpendiculars to determine F, but we may use the third to check the precision of our drawing.) And what remains to do is easy. Let M be the point of intersection of D_1F and BC (see

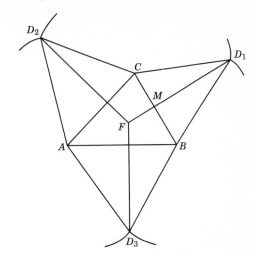

Fig. 14.3. The common destination of three travelers.

Fig. 14.3). Construct the right triangle *FMD* (see Fig. 14.4), with hypotenuse $MD = MD_1$ and leg *MF*. Obviously, *FD* is the altitude [our supplementary aim (a)] and $\angle FMD$ measures the dihedral angle included by the base $\triangle ABC$ and the lateral face $\triangle DBC$ which was required by our problem.

(8) One of the virtues of a good problem is that it generates other good problems.

The foregoing solution may, and should, leave a doubt in our mind. We found the result represented by Fig. 14.3 (that the three perpendiculars described above are concurrent) by considering the motion of rotating bodies. Yet the result is a proposition of geometry and so it should be established independently of the idea of motion, by geometry alone.

Now, it is relatively easy to free the foregoing consideration [in subsections (6) and (7)] from ideas of motion and establish the result by

Fig. 14.4. The rest is easy.

ideas of solid geometry (intersection of spheres, orthogonal projection).[3] Yet the result is a proposition of plane geometry and so it should be established independently of the ideas of solid geometry, by plane geometry alone. (How?)

(9) Observe that this tenth grade problem also illustrates various points about teaching discussed in the foregoing. For instance, the students could and should participate in the final formulation of the problem, there is an exploratory phase, and a rich background.

Yet here is the point I wish to emphasize: the problem is designed to deserve the attention of the students. Although the problem is not so close to everyday experience as our seventh grade problem, it starts from a most familiar piece of knowledge (the construction of a triangle from three sides), it stresses from the start an idea of general interest (analogy), and it points to eventual practical applications (mechanical drawing). With a little skill and good will, the teacher should be able to secure for this problem the attention of all students who are not hopelessly dull.

14.7. Learning teaching

There remains one more topic to discuss and it is an important topic: teacher training. In discussing this topic, I am in a comfortable position: I can almost agree with the "official" standpoint. (I am referring here to the "Recommendations of the Mathematical Association of America for the training of mathematics teachers," *American Mathematical Monthly,* **67** (1960) 982–991. Just for the sake of brevity, I take the liberty to quote this document as the "official recommendations.") I shall concentrate on just two points. To these two points I have devoted a good deal of work and thought in the past and practically all my teaching in the last ten years.

To state it roughly, one of the two points I have in mind is concerned with "subject matter" courses, the other with "methods" courses.

(1) *Subject matter.* It is a sad fact, but by now widely recognized, that our high school mathematics teachers' knowledge of their subject matter is, on the average, insufficient. There are, certainly, some well-prepared high school teachers, but there are others (I met with several) whose good will I must admire but whose mathematical preparation is not admirable. The official recommendations of subject matter courses may not be perfect, but there is no doubt that their acceptance would result in substantial improvement. I wish to direct your attention to a point which, in my considered opinion, should be added to the official recommendations.

---

[3] *Cf.* sect. 6.2(3).

Our knowledge about any subject consists of information and know-how. Know-how is ability to use information; of course, there is no know-how without some independent thinking, originality, and creativity. Know-how in mathematics is the ability to do problems, to find proofs, to criticize arguments, to use mathematical language with some fluency, to recognize mathematical concepts in concrete situations.

Everybody agrees that, in mathematics, know-how is more important, or even much more important, than mere possession of information. Everybody demands that the high school should impart to the students not only information in mathematics but know-how, independence, originality, creativity. Yet almost nobody asks these beautiful things for the mathematics teacher—is it not remarkable? The official recommendations are silent about the mathematical know-how of the teacher. The student of mathematics who works for a Ph.D. degree must do research, yet even before he reaches that stage he may find some opportunity for independent work in seminars, problem seminars, or in the preparation of a master's thesis. Yet no such opportunity is offered to the prospective mathematics teacher—there is no word about any sort of independent work or research work in the official recommendations. If, however, the teacher has had no experience in creative work of some sort, how will he be able to inspire, to lead, to help, or even to recognize the creative activity of his students? A teacher who acquired whatever he knows in mathematics purely receptively can hardly promote the active learning of his students. A teacher who never had a bright idea in his life will probably reprimand a student who has one instead of encouraging him.

Here, in my opinion, is the worst gap in the subject matter knowledge of the average high school teacher: he has no experience of active mathematical work and, therefore, he has no real mastery even of the high school material he is supposed to teach.

I have no panacea to offer, but I have tried one thing. I have introduced and repeatedly conducted a *problem solving seminar* for teachers. The problems offered in this seminar do not require much knowledge beyond the high school level, but they require some degree, and now and then a higher degree, of concentration and judgment—and, to that degree, their solution is "creative" work. I have tried to arrange my seminar so that the students should be able to use much of the material offered in their classes without much change; that they should acquire some mastery of high school mathematics; and so that they should have even some opportunity for practice teaching (in teaching each other in small groups). *Cf.* vol. 1, pp. 210–212.

(2) *Methods*. From my contact with hundreds of mathematics teachers I gained the impression that "methods" courses are often re-

ceived with something less than enthusiasm. Yet so also are received, by the teachers, the usual courses offered by the mathematics departments. A teacher with whom I had a heart to heart talk about these matters found a picturesque expression for a rather widespread feeling: "The mathematics department offers us tough steak which we cannot chew and the school of education vapid soup with no meat in it."

In fact, we should once summon up some courage and discuss publicly the question: Are methods courses really necessary? Are they in any way useful? There is more chance to reach the right answer in open discussion than by widespread grumbling.

There are certainly enough pertinent questions. Is teaching teachable? (Teaching is an art, as many of us think—is an art teachable?) Is there such a thing as the teaching method? (What the teacher teaches is never better than what the teacher is—teaching depends on the whole personality of the teacher—there are as many good methods as there are good teachers.) The time allotted to the training of teachers is divided between subject matter courses, methods courses, and practice teaching; should we spend less time on methods courses? (Many European countries spend much less time.)

I hope that people younger and more vigorous than myself will take up these questions some day and discuss them with an open mind and pertinent data.

I am speaking here only about my own experience and my own opinions. In fact I have already implicitly answered the main question raised; I believe that methods courses may be useful. In fact, what I have presented in the foregoing was a sample of a methods course, or rather an outline of some topics which, in my opinion, a methods course offered to mathematics teachers should cover.

In fact, all the classes I have given to mathematics teachers were intended to be methods courses to some extent. The name of the class mentioned some subject matter, and the time was actually divided between that subject matter and methods: perhaps nine tenths for subject matter and one tenth for methods. If possible, the class was conducted in dialogue form. Some methodical remarks were injected incidentally, by myself or by the audience. Yet the derivation of a fact or the solution of a problem was almost regularly followed by a short discussion of its pedagogical implications. "Could you use this in your classes?" I asked the audience. "At which stage of the curriculum could you use it? Which point needs particular care? How would you try to get it across?" And questions of this nature (appropriately specified) were regularly proposed also in examination papers. My main work, however, was to choose such problems (like the two problems I have here presented) as would illustrate strikingly some pattern of teaching.

(3) The official recommendations call "methods" courses "curriculum-study" courses and are not very eloquent about them. Yet you can find there one excellent recommendation, I think. It is somewhat concealed; you must put two and two together, combining the last sentence in "curriculum study courses" and the recommendations for Level IV. But it is clear enough: A college instructor who offers a methods course to mathematics teachers should know mathematics at least on the level of a Master's degree. I would like to add: he should also have had some experience, however modest, of mathematical research. If he had no such experience how could he convey what may be the most important thing for prospective teachers, the spirit of creative work?

You have now listened long enough to the reminiscences of an old man. Some concrete good could come out of this talk if you give some thought to the following proposal which results from the foregoing discussion. I propose that the following two points should be added to the official recommendations of the Association:

I. *The training of teachers of mathematics should offer experience in independent ("creative") work on the appropriate level in the form of a Problem Solving Seminar or in any other suitable form.*

II. *Methods courses should be offered only in close connection either with subject matter courses or with practice teaching and, if feasible, only by instructors experienced both in mathematical research and in teaching.*

14.8. The teacher's attitude[4]

As I have already said, my classes addressed to teachers were, to some extent, "methods" courses. In these classes I aimed at points of immediate practical use in the daily task of the teacher. Therefore, inavoidably, I had to express repeatedly my views on the teacher's daily task and on the teacher's mental attitude. My comments tended to assume a set form and eventually I was led to condense them into "Ten Commandments for Teachers"; see p. 116. I wish to add a few comments on these ten rules.

In formulating these rules, I had in mind the participants in my classes, teachers who teach mathematics on the high school level. Nevertheless, these rules are applicable to any teaching situation, to any subject taught on any level. Especially on the high school level, however, the mathematics teacher has more and better opportunities

---

[4] This section can be read independently of the foregoing (of which a few points will be repeated.) It is reprinted, with several modifications and the kind permission of the editor, from the *Journal of Education of the Faculty and College of Education, Vancouver and Victoria*, no. 3, 1959, pp. 61–69.

TEN COMMANDMENTS FOR TEACHERS

1. Be interested in your subject.
2. Know your subject.
3. Know about the ways of learning: The best way to learn anything is to discover it by yourself.
4. Try to read the faces of your students, try to see their expectations and difficulties, put yourself in their place.
5. Give them not only information, but "know-how," attitudes of mind, the habit of methodical work.
6. Let them learn guessing.
7. Let them learn proving.
8. Look out for such features of the problem at hand as may be useful in solving the problems to come—try to disclose the general pattern that lies behind the present concrete situation.
9. Do not give away your whole secret at once—let the students guess before you tell it—let them find out by themselves as much as is feasible.
10. Suggest it, do not force it down their throats.

to apply some of them than the teacher of other subjects; and this refers in particular to rules 6, 7, and 8.

On what authority are these commandments founded? Dear fellow teacher, do not accept any authority except your own well-digested experience and your own well-considered judgement. Try to see clearly what the advice means in your particular situation, try the advice in your classes, and judge after a fair trial.

Let us now consider the ten rules one by one, with especial attention to the task of the mathematics teacher.

(1) There is just one infallible teaching method: if the teacher is bored by his subject, his whole class will be infallibly bored by it.

This should be enough to render evident the first and foremost commandment for teachers: *Be interested in your subject.*

(2) If a subject has no interest for you, do not teach it, because you will not be able to teach it acceptably. Interest is a *sine qua non,* an indispensably necessary condition; but, in itself, it is not a sufficient condition. No amount of interest, or teaching methods, or whatever else will enable you to explain clearly a point to your students that you do not understand clearly yourself.

This should be enough to render obvious the second commandment for teachers: *Know your subject.*

Both interest in, and knowledge of, the subject matter are necessary for the teacher. I put interest first because with genuine interest you have a good chance to acquire the necessary knowledge, whereas some knowledge coupled with lack of interest can easily make you an exceptionally bad teacher.

(3) You may benefit a great deal from reading a good book or listening to a good lecture on the psychology of learning. Yet reading and listening are not absolutely necessary, and they are by no means sufficient; you should *know the ways of learning,* you should be intimately acquainted with the process of learning from *experience*—from the experience of your own studies and from the observation of your students.

Accepting hearsay as evidence for a principle is bad; paying lip service to a principle is worse. Now, there is one case where you cannot afford to be content with hearsay and lip service, there is one principle of learning that you should earnestly realize: the principle of active learning.[5] Try to see at least its central point: *The best way to learn anything is to discover it by yourself.*

(4) Even with some genuine knowledge and interest and some understanding of the learning process you may be a poor teacher. The case is unusual, I admit, but not quite rare; some of us have met with an otherwise quite competent teacher who was unable to establish "contact" with his class. In order that teaching by one should result in learning by the other, there must be some sort of contact or connection between teacher and student: the teacher should be able to see the student's position; he should be able to espouse the student's cause. Hence the next commandment: *Try to read the faces of your students, try to see their expectations and difficulties, put yourself in their place.*

The response of the students to your teaching depends on their background, their outlook, and their interests. Therefore, keep in mind and take into account what they know and what they do not know, what they would like to know and what they do not care to know, what they ought to know and what is less important for them to know.

(5) The four foregoing rules contain the essentials of good teaching. They form jointly a sort of necessary and sufficient condition: if you have interest in, and knowledge of, the subject matter and if, moreover, you can see the student's case and what helps or hampers his learning, you are already a good teacher or you will become one soon; you may need only some experience.

It remains to spell out some consequences of the foregoing rules,

---

[5] See sect. 14.4(1), 14.5(1). To be acquainted with the two other principles discussed in the foregoing is advisable.

especially such consequences as concern the high school mathematics teacher.

Knowledge consists partly of "information" and partly of "know-how." Know-how is skill; it is the ability to deal with information, to use it for a given purpose; know-how may be described as a bunch of appropriate mental attitudes; know-how is ultimately the ability to work methodically.

In mathematics, know-how is the ability to solve problems, to construct demonstrations, and to examine critically solutions and demonstrations. And, in mathematics, know-how is much more important than the mere possession of information. Therefore, the following commandment is of especial importance for the mathematics teacher: *Give your students not only information, but know-how, attitudes of mind, the habit of methodical work.*

Since know-how is more important in mathematics than information, it may be more important in the mathematics class how you teach than what you teach.

(6) First guess, then prove—so does discovery proceed in most cases. You should know this (from your own experience, if possible), and you should know, too, that the mathematics teacher has excellent opportunities to show the role of guessing in discovery and thus to impress on his students a fundamentally important attitude of mind. This latter point is not so widely known as it should be and, just for this reason, it deserves particular attention. I wish you would not neglect your students in this respect: *Let them learn guessing.*

Ignorant and careless students are liable to come forward with "wild" guesses. What we have to teach is, of course, not wild guessing, but "educated" "reasonable" guessing. Reasonable guessing is based on judicious use of inductive evidence and analogy, and ultimately encompasses all procedures of *plausible reasoning* which play a role in "scientific method."[6]

(7) "Mathematics is a good school of plausible reasoning." This statement summarizes the opinion underlying the foregoing rule; it sounds unfamiliar and is of very recent origin; in fact, I think that I should claim credit for it.

"Mathematics is a good school of demonstrative reasoning." This statement sounds very familiar—some form of it is probably almost as old as mathematics itself. In fact, much more is true: mathematics is coextensive with demonstrative reasoning, which pervades the sciences just as far as their concepts are raised to a sufficiently abstract and definite, mathematico-logical level. Under this high level there is no

---

[6] See chapter 15.

place for truly demonstrative reasoning (which is out of place, for instance, in everyday affairs). Still (it is needless to argue such a widely accepted point), the mathematics teacher should acquaint all his students beyond the most elementary grades with demonstrative reasoning: *Let them learn proving.*

(8) Know-how is the more valuable part of mathematical knowledge, much more valuable than the mere possession of information. Yet how should we teach know-how? The students can learn it only by imitation and practice.

When you present the solution of a problem, *emphasize* suitably the *instructive features* of the solution. A feature is instructive if it deserves imitation; that is, if it can be used not only in the solution of the present problem, but also in the solution of other problems—the more often usable, the more instructive. Emphasize the instructive features not just by praising them (which could have the contrary effect with some students) but by your *behavior* (a bit of acting is very good if you have a bit of theatrical talent). A well-emphasized feature may convert your solution into a *model solution,* into an impressive *pattern* by imitating which the student will solve many other problems. Hence the rule: *Look out for such features of the problem at hand as may be useful in solving the problems to come—try to disclose the general pattern that lies behind the present concrete situation.*[7]

(9) I wish to indicate here a little classroom trick which is easy to learn and which every teacher should know. When you start discussing a problem, let your students guess the solution. The student who has conceived a guess, or has even stated his guess, commits himself: he has to follow the development of the solution to see whether his guess comes true or not—and so he cannot remain inattentive.[8]

This is just a very special case of the following rule, which itself is contained in, and spells out, some parts of rules 3 and 6: *Do not give away your whole secret at once—let the students guess before you tell it— let them find out by themselves as much as is feasible.*

In fact, the credit for this rule is due to Voltaire who expressed it more wittily: "Le secret d'être ennuyeux c'est de tout dire." "The art of being a bore consists in telling everything."

(10) A student presents a long computation which goes through several lines. Looking at the last line, I see that the computation is wrong, but I refrain from saying so. I prefer to go through the computation with the student, line by line: "You started out all right, your first line is correct. Your next line is correct too; you did this and that.

---

[7] Do you want more details? Read the whole book.
[8] *Cf.* sect. 14.5(2).

The next line is good. Now, what do you think about this line?" The mistake is on that line and if the student discovers it by himself, he has a chance to learn something. If, however, I at once say "This is wrong" the student may be offended and then he will not listen to anything I may say afterwards. And if I say "This is wrong" once too often, the student will hate me and mathematics and all my efforts will be lost as far as he is concerned.

Dear fellow teacher, avoid saying "You are wrong." Say instead, if possible: "You are right, but. . . ." If you proceed so, you are not hypocritical, you are just humane. That you should proceed so, is implicitly contained in rule 3. Yet we can render the advice more explicit: *Suggest it, do not force it down their throats.*

Our last two rules, 9 and 10, tend in the same direction. What they jointly suggest is to leave the students as much freedom and initiative as possible under existing teaching conditions. Pressed for time, the mathematics teacher is often tempted to sin against the spirit of these rules, the *principle of active learning.* He may hurry to the solution of a problem without leaving enough time for the students to put the problem to themselves in earnest. He may name a concept or formulate a rule too soon, without sufficient preparation by appropriate material, before the students can feel the need for such a concept or rule. He may commit the celebrated mistake of *deus ex machina:* he may introduce some device (for instance, a tricky auxiliary line in a geometric proof) which leads to the result all right, but the students cannot see for their life how it was humanly possible to discover such a trick which appeared right out of the blue.

There are too many temptations to violate the principle. Let us, therefore, emphasize a few more of its facets.

Let your students ask the questions; or ask such questions as they may ask by themselves.

Let your students give the answers; or give such answers as they may give by themselves.

At any rate avoid answering questions that nobody has asked, not even yourself.

Examples and Comments on Chapter 14

First Part

14.1. Taking $122°25'41''$ west for the longitude of San Francisco answer the question (d) of sect. 14.6(1).

14.2. Following up a hint in sect. 14.6(8), prove the proposition suggested by Fig. 14.3 by solid geometry.

14.3. (Continued) Prove the proposition suggested by plane geometry.

14.4. Sect. 14.6(9) mentions some points discussed in the foregoing and illustrated by the problem presented in sect. 14.6(3) to 14.6(8). Do you notice more such points?

Second Part

14.5. *Why problem solving?* I am of the opinion that teaching "problem solving" could be an essential ingredient of various curricula rather different in other respects and should be, in fact, an essential ingredient of any useful high school mathematics curriculum. This opinion underlies the present book and my related writings and has already been explicitly stated in the foregoing (see sect. 5 of the preface and sect. 14.2). If the reader is not convinced by the foregoing chapters that this opinion has some merit, I cannot do much for him. Still, I wish to present a few comments on the role of problem solving in the high school curriculum.

(1) We are concerned here with the teaching of mathematics on the high school level and with the aims of such teaching. Responsible and realistic consideration of these aims should take into account the use that the students can be expected to make of what they are supposed to learn. Of course, there are different categories of students, and some will make more, and others less, use of the knowledge acquired in school, and some categories form a greater, and others a smaller, fraction of the student body. Reliable statistical data about these things would be very desirable, but they are scarcely available. The numerical proportions I shall use in what follows are rough estimates without serious statistical basis—I use them, in fact, just for the sake of concreteness.

(2) Let us consider such students as take some mathematics on the high school level (algebra, geometry, etc.) and, with regard to the future use of such study in their respective professions, let us distinguish three categories: mathematicians, users of mathematics, and nonusers of mathematics.

Let us draw the boundaries of the first category rather widely: Let us count as "mathematicians" or "producers of mathematics" also theoretical physicists, astronomers, and certain engineers in special research positions. Altogether they may form about 1% of the students. (The number of future Ph.D.'s in mathematics is closer to 0.1%.)

Engineers, scientists (also some social scientists), mathematics and science teachers, etc. are users (but, in general, not producers) of mathematics. Let us count as users of mathematics also such students as will not use mathematics in their profession, but who essentially need some mathematics in their studies (such is the case of numerous engineering graduates who become salesmen or managers). The number of all sorts of users of mathematics may be, say, 29% of the students.

Many of the remaining students could, but actually will not, use any mathematics beyond what they should have learned in primary school. It is a rough but not unrealistic estimate that 70% of the students will be nonusers of mathematics; almost all future businessmen, lawyers, clergymen, etc. belong to this category.

(3) We do not know in advance who will become what and so we do not know which students belong to which category. The mathematics class, therefore, should be so conducted that it conforms to two "principles":

First, each student should be able to derive some profit from his study irrespective of future occupation.

Second, such students as have some aptitude for mathematics should be attracted to it and not get disgusted with it by ill-advised teaching.

I take for granted that the reader accepts these two principles, at least to some extent. In fact, I think that to plan the curriculum without continuous earnest attention to these two principles would be unrealistic and irresponsible.

Let me hint briefly how the three categories of students here considered [see (2)] could gain something essential from the study of problem solving.

(4) The ability to solve mathematical problems needs, of course, some knowledge of the mathematical subject matter involved, but, moreover, it needs certain useful habits of mind, a certain general attitude which we are inclined to call "common sense" in everyday life. The teacher who wishes to serve equally all his students, future users and nonusers of mathematics, should teach problem solving so that it is about one-third mathematics and two-thirds common sense. It may not be too easy to implant common sense and useful mental habits, but if the mathematics teacher succeeds in implanting them he has rendered real service to his students, whatever their future occupation may be. This service is certainly the most significant thing he can do for those 70% of students who will have no use for technical mathematics in later life.

The 29% of students who will become users of mathematics need, as a preparation for subsequent studies, some technical proficiency (for instance, some facility in algebraic manipulation). Yet just the students with a good practical mind are reluctant to learn technicalities unless they are convinced that those technicalities serve a purpose, are good for something. The best the teacher can do to justify the teaching of technicalities is to demonstrate that they are efficacious in solving naturally arising interesting concrete problems.

Future mathematicians form only about 1% of the student body, but it is of paramount importance that they should be discovered: if they choose the wrong profession, their talent, needed by modern society in more than one way, may be wasted. The most important thing the high school teacher can do for this 1% is to awake their interest in mathematics. (To teach them a little more or less subject matter in high school hardly matters as, in any case, it can be only an infinitesimal fraction of the subject matter they have to know ultimately.) Now, problem solving is an important avenue to mathematics; it is not the only one, but it is connected with other important avenues (see ex. 14.6). Moreover, the teacher should treat a few problems which, although somewhat more difficult and time consuming, have real mathematical beauty and background (see chapter 15).

(5) I hope, as I said before, that arguments for teaching problem solving in the high school can be found in every part of the present book and of my related writings. A few specific points will be emphasized in the following.

14.6. *Problem solving and theory formation.* A devoted and well-prepared teacher can take a significant but not too complex problem and by helping the students discover its various aspects he can lead them through it, as through a gateway, into a whole theory. To prove that $\sqrt{2}$ is irrational, or that there are infinitely many prime numbers, are such significant problems. The former can be a gateway to a critical concept of the real number,[9] the latter to number theory.[10]

Such a procedure of the teacher parallels the historical evolution of science. The solution of a significant problem, the effort expended on, and the insight gained from, the solution may open the gate to a new science, or even to a new era of science. We should think of Galileo and the problem of falling bodies, or of Kepler and the problem of the orbit of Mars.

In his work quoted, M. Wagenschein offers an idea which deserves, in my opinion, the attention of all planners of curricula: instead of hurrying through all the details of a much too extended program, the teacher should concentrate on a few really significant problems and treat them leisurely and thoroughly. The students should explore all aspects of the problem accessible at their level, they should discover the solution by themselves, they should anticipate, led by the teacher, some consequences of the solution. In this way, a problem may become a representative example, a paradigm for a whole chapter of science. This is a first sketch of the idea of *paradigmatic teaching* about which every teacher seriously concerned with the curriculum should read more in the book quoted; see also ex. 14.11.

Let us note that the appropriate treatment of a single problem may be the gateway to, or the representative example of, a branch of science. It is in view of this and similar observations that I took the liberty to say in sect. 14.2 "thinking may be identified, at least in first approximation, with problem solving."

14.7. *Problem solving and general culture.* Many people think (and I am one of those many) that one of the essential tasks, and perhaps the most essential task, of the high school is to impart general culture. Let us not enter here, however, upon defining "general culture," because otherwise we could wrangle indefinitely about the "right" definition.

In teaching problem solving in the mathematics class we have an excellent opportunity to develop certain concepts and habits of mind which are, in my opinion, important ingredients of general culture; the box on p. 124 displays a list.[11] The list is not exhaustive; it contains only such more conspicuous, more tangible points as are, I hope, not beyond the reach of an average high school class. Most items of the list are explained at length in this book and in my related writings.[12] For additional remarks see ex. 14.8 and 14.10.

For a viewpoint which raises important questions about the relation between general culture and the teaching of mathematics see the work of Wittenberg, footnote 9.

---

[9] *Cf.* A. I. Wittenberg, Bildung und Mathematik (Klett Verlag 1963) pp. 168–253 *passim.*

[10] *Cf.* Martin Wagenschein, Exemplarisches Lehren im Mathematikunterricht, *Der Mathematikunterricht,* vol. 8, 1962, part 4, pp. 29–38.

[11] From *American Mathematical Monthly,* vol. 65, 1958, p. 103.

[12] For "Language of Formulas" see chapter 2, for "Reasonable Guessing" chapter 15. For Unknown, Data, and Condition, Generalization, Specialization, and Analogy, see the Index.

Unknown	Data	Condition

Generalization   Specialization

Analogy

Strict Reasoning

Reasonable Guessing

Language

of

Diagrams   Formulas

14.8. *Language of figures.* There are people who have an urge to represent their ideas by means of some sort of geometric symbols. In their mind some figures of speech which we all use have a tendency to develop into geometric figures. When they are thinking about their problems they are apt to find paper and pencil and start doodling: They may be struggling to express themselves in a language of geometric figures.

(1) There are many important nongeometric facts and ideas which are most appropriately expressed by geometric figures, graphs, or diagrams. By dots placed at appropriate elevations, high or low, the musical notation expresses the pitch of sounds, high or low. By means of geometric symbols (dots and connecting strokes) the chemical notation expresses the constitution of chemical compounds. Numbers and numerical relations can be expressed in various ways by geometrical objects and relations. Analytic geometry offers systematic means of translating numerical relations into geometrical relations and vice versa. Analytic geometry is, in a way, a dictionary of two languages, of the language of formulas and of the language of geometric figures: it enables us to translate readily from one language into the other. The ideas of analytic geometry underlie a great variety of graphs, diagrams, nomograms, etc. used in science, engineering, economics, and so on. Diagrams may also be useful in pure mathematics, and some can be well explained on the high school level; for an illustration not found in the usual textbooks, see ex. 14.9.

(2) Graphs and diagrams used in science are usually definite and precise in principle: the (idealized) geometric figure represents exactly the intended numerical relations. It is important to observe that graphical representations which are more or less vague or unprecise may still be useful. For instance, I believe that Fig. 11.2 has some suggestive value although it is scarcely more than a metaphor on paper, a figure of speech changed into a visible figure; Fig. 11.3 and 11.4 are of similar nature. Figures 15.1 to 15.5 have a clear mathematical meaning: they represent the set of all triangular shapes and certain subsets of this set. Yet their principal interest is that they hint something more, a process which we cannot yet conceive so clearly, the progress of an inductive argument.

Of two diagrams similar to the eye, one may be interpreted in an entirely definite mathematical meaning and the other in a rather vague metaphorical sense, and all gradations are possible between strict precision and poetical allusion. Flow charts, used for many purposes, can well illustrate the point.

(3) Geometry, our knowledge of space, has several aspects. Geometry can be conceived, as we know, as a science based on axioms. Yet geometry is also a skill of the eyes and the hands. Again, geometry can be considered as a part of physics (the most primitive part, say some physicists—the most interesting part, say some geometers). As a part of physics, geometry is also a field in which we can make intuitive or inductive discoveries and verify them subsequently by reasoning. To these aspects the foregoing has added one more: geometry is also the source of the symbols of a sort of language which can be colloquial or precise, and both ways helpful and enlightening.

There is a moral for the teacher: if you wish to instruct your students and not just hurry through the items of a curriculum dictated from above, do not neglect any of these aspects. Especially, do not insist too early or too much on the axiomatic aspect of geometry if you do not wish to disgust the future engineers and scientists among your students (or future artists and philosophers) who may be more attracted by the eye-knowledge of geometrical shapes, by space visualization, or by inductive discovery, or the powerful help to thinking afforded by diagrammatic representation.

14.9. *Rationals and irrationals.* What follows is only a rapid sketch of what should be very carefully done in the classroom—we are facing here what is perhaps the most delicate point of the high school mathematics curriculum. For the sake of brevity, I shall use a few terms and the notation of analytic geometry of which, however, very little knowledge is actually needed (just a little "graphing").

Let x and y denote, as usual, rectangular coordinates. The line with the equation $y = 1$ is the *number line* (just a "glorified yardstick"). See Fig. 14.5, which also displays the *lattice points,* that is, the points with integral coordinates. Figure 14.5 emphasises the lattice points along the number line ("milestones along a long straight road").

In Fig. 14.5 the number x is represented by the point $(x, 1)$ on the number line. We draw the straight line through the point $(x, 1)$ and the origin $(0, 0)$. Iff this

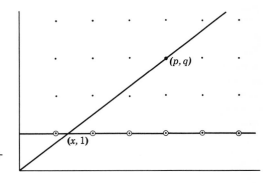

Fig. 14.5. Number line and lattice points.

line passes through the lattice point (p, q) (different from the origin!) the number x is rational:

$$\frac{x}{1} = \frac{p}{q}$$

from similar triangles.

The teacher should ask (but *not answer*, at least not answer for a considerable time) the following questions:

The origin $(0, 0)$ is a lattice point. Must every straight line passing through the origin pass also through another lattice point?

Of course, there are just two cases: a straight line through the origin does or does not pass through another lattice point; which case is more *likely?*

The teacher should let these questions and Fig. 14.5 "sink in" and only when the students have realized their import (perhaps after hours or weeks or months) should he start discussing the irrationality of $\sqrt{2}$, the approximation of irrationals by rationals (Fig. 14.5 may be used, after Felix Klein, as a gateway to continued fractions), etc.

14.10. *Strict reasoning.* Should we teach mathematical proofs in the high school? In my opinion, there is little doubt about the answer: Yes, we should, unless extremely adverse conditions compel us to lower the standard. Rigorous proofs are the hallmark of mathematics, they are an essential part of mathematics' contribution to general culture. The student who was never impressed by a mathematical proof missed a basic mental experience.

On what level of rigor should we teach mathematical proofs, and how? The answer to this question is not so simple; it is, in fact, beset with difficulties. Yet to ignore these difficulties and give some answer without much thought, just by following tradition, fashion, or prejudice, is not the kind of thing conscientious planners of the curriculum should do.

There are proofs and proofs, there are various ways of proving. The first thing we must realize is that certain ways of proving are more appropriately taught on a given level of age and maturity than other ways.

(1) A certain aspect of mathematical proofs was observed and described by Descartes with remarkable lucidity.

I quote the third of his *Rules for the Direction of the Mind:*[13] "About the objects of our study, we should not seek the opinion of others or our own conjectures, but only what we can see with clear and evident intuition or deduce with certitude, for there are no other ways to knowledge." In explaining this rule, Descartes considers the "two ways to knowledge," intuition and deduction, successively. Here is the beginning of his discussion of deduction:[14] "The evidence and certitude of intuition are required not only in propositions, but also in all kinds of reasoning. For instance, we intend to deduce that 2 and 2 equals 3 and 1. Then we must intuitively see not only that 2 and 2 make 4 and that 3 and 1 make 4, but also that from these two propositions that third one mentioned above necessarily follows."

---

[13] *Oeuvres,* vol. X, p. 366.
[14] *Oeuvres,* vol. X, p. 369.

A mathematical deduction appears to Descartes as a chain of conclusions, a sequence of successive steps. What is needed for the validity of deduction is *intuitive insight* at each step which shows that the conclusion attained by that step evidently flows and necessarily follows from formerly acquired knowledge (acquired directly by intuition or indirectly by previous steps of deduction). (We know from chapter 7 that a graph with ramifications represents the structure of a proof more appropriately than a simple unramified chain, but this is not the point: If Descartes knew the diagrammatic representation we have studied in chapter 7, he would insist that each element of the diagram—such as represented by Fig. 7.18—must be supported by intuitive evidence.)

(2) Yet mathematics has several aspects. It can also be regarded as a game with symbols played according to arbitrarily fixed rules in which the principal consideration is to stick to the rules of the game. (This aspect is quite recent; fifty years ago most mathematicians and most philosophers would have thought that such an aspect of mathematics is revolting. Still, this aspect, which has been introduced under the influence of the great mathematician David Hilbert, is quite appropriate in certain studies of the foundations of mathematics.)

In this game with symbols, the symbols are meaningless (we are supposed to ignore their meaning if they have one). There are "proofs" in this game; a step of a proof consists in writing down a "well-formed" new formula (a combination of symbols complying with the rules). Such a step is considered as valid if the new formula is written down strictly in accordance with certain formulas introduced at the beginning ("axioms") with formulas written in previous steps, and with certain rules of inference fixed from the start. To be handled in this fashion, both the proofs and the propositions proved must be "atomized," decomposed into very small steps and minute component parts.

(3) Between the two extreme aspects considered in (1) and (2) there are others.[15] In fact, the concept of mathematical proof has evolved, changed from one era of science to the other. The study of this evolution and of the motives that led to it could have great interest for us teachers. By understanding how the human race has acquired a concept we could better see how the human child should acquire it. *Cf.* ex. 14.13.

A productive mathematician is, of course, free to prefer any aspect of mathematics; he should prefer the aspect most profitable for his work. Yet on the high school level our choice is not free, and if the choice is between (1) and (2) (between teaching proofs closer to one aspect or to the other), we can hardly hesitate.

I think that everybody prefers intuitive insight to formal logical arguments, including professional mathematicians. Jacques Hadamard, an eminent French mathematician of our times, expressed it so:[16] "The object of mathematical rigor is to sanction and legitimize the conquests of intuition, and there never was any other object for it." Yet, if we exclude professional mathematicians, almost nobody remains who would be in a position to properly appreciate formal argu-

---

[15] For a well-illustrated important investigation of the nature of proofs see Lakatos (quoted in the Bibliography).

[16] Émile Borel, *Lecons sur la théorie des fonctions,* 3rd ed. 1928; see p. 175.

ments. Intuition comes to us "naturally," formal arguments do not.[17] At any rate, intuition comes to us much earlier and with much less outside influence than formal arguments which we cannot really understand unless we have reached a relatively high level of logical experience and sophistication.

Therefore, I think that in teaching high school age youngsters we should emphasize intuitive insight more than, and long before, deductive reasoning. And when we present proofs, we should present them much closer to the idea of Descartes [see (1)] than to a certain idea of certain modern logicians [see (2)].

I have met with youngsters who had a definite interest, and probably some talent, for engineering or science, but refused to learn mathematics, and I have a hunch where this refusal comes from.

(4) Let me give an example. I consider the proposition: *Among three points on a straight line, there is just one that lies between the two others.*

Observe that this proposition says something essential about the nature of the straight line. If there are three points on a circle, no one plays a special role, no one is distinguished from the others by "betweenness."

Does this proposition about the three points on the straight line *need a proof?* In a university lecture on the foundations of geometry, a proof for this proposition starting from the axioms may be essential. Yet to present such a proof to a tenth grade high school class just starting the rational study of geometry is simply preposterous.

This is my opinion, which may be wrong. To have a defensible opinion you must picture to yourself the reaction of the class to such a proof. I imagine it so: The majority of the youngsters will be simply bored. Yet a less mediocre and less indifferent minority will feel more or less distinctly that the proof is superfluous and aimless. There may be one or two boys in the class who will be frankly disgusted and revolted. At any rate, I think that such would have been my own reaction if such a proof had been presented to me when I was of high school age. I do not pretend that I remember exactly the ideas of the teenager I was sixty years ago, and I certainly do not pretend that that teenager was always right. Still, I can vividly imagine my reaction to such a proof: It would have convinced me that my teacher is stupid—or that mathematics is stupid, or that both are stupid. And, at that point, I would have stopped listening to the explanations of the teacher—or, if obliged to listen, I would have listened with reluctance, suspicion, and contempt.

At any rate, I think that an adverse reaction to a proof of the nature outlined is natural and proper.

(5) There are many aspects of proofs. I think that the role of proofs in building up science is more complex than it is usually assumed, and there may be a question deserving a philosophical inquiry. We deal here, however, with another question: which aspect of proofs should be presented to beginners? This question seems to me easier, and I have a pretty firm opinion about it which I take the liberty to express.

In the first place, the beginner must be convinced that proofs deserve to be studied, that they have a purpose, that they are interesting.

---

[17] For an incidental remark expressing a very similar opinion see H. Weyl, *Philosophy of mathematics and natural science*, p. 19.

Proofs have a purpose in a law court. The defendant is suspected to be guilty, but this is just suspicion, there is a doubt. It must be proved either that he is guilty or that he is not. The purpose of a legal proof is to *remove a doubt,* but this is also the most obvious and natural purpose of a mathematical proof. We are in doubt about a clearly stated mathematical assertion, we do not know whether it is true or false. Then we have a problem: to remove the doubt, we should either prove that assertion or disprove it.

Now I can explain why I am so firmly convinced that the proof hinted at previously (about the three points on a line) is out of place in the high school. A high school age youngster who has understood that assertion about the three points can hardly doubt it. There is no doubt to remove, and so the proof appears useless, aimless, senseless. The case is aggravated if the proof starts from axioms, distinguishes several cases, and takes thirteen lines in the textbook. It may give the youngsters the impression that mathematics consists in proving the most obvious things in the least obvious way.

(6) Yet, as I have already said, the proof about the three points on the straight line is quite in order on the proper level. When we teach it in high school we commit the ugly and ridiculous pedagogical sin of the *confusion of levels.* See ex. 14.15.

On the research level, it can happen that a proposition appears intuitively obvious, we have strong plausible arguments for it, but no formal argument. It is in such a situation that a mathematician may do his best to discover a proof. For a preview of such situations on the high school level see ex. 14.11, and more examples in chapter 15.

(7) Before leaving this subject, I must sound a warning against another grave pedagogical error: overemphasizing trivial proofs. Crowding the pages of a textbook with pointless proofs which lack motivation and rewarding goals may make the worst impression on the best students, who have some gift of intuition which could be most useful in engineering or science or mathematics.

Also this ugly error may be due to a confusion of levels. Not for a high school age boy, but for a professional mathematician it may be necessary to check the formal justification for each step of a long argument. Such checking may be necessary, although it is not the most enjoyable part of the mathematician's work. Logic is the lady at the exit of the supermarket who checks the price of each item in a large basket whose contents she did not collect.

14.11. *Can a map be perfect?* A map is the representation of a part of the earth's surface on a flat piece of paper.

(1) To understand the situation before us, we generalize it and we describe more precisely the more general situation. (This transition from the particular to the general and from the more intuitive to the more abstract level is important; it is made here abruptly, but in the classroom it should be made gradually and carefully).

We consider the mapping of a surface S onto another surface S'. We consider a *one-to-one mapping,* that is, we suppose that to each point p of S there corresponds just one point p' of S', the *image* of p, and that, conversely, to each point of p' of S' there corresponds just one point p of S, the *original* of p'. We assume further that the mapping is "continuous": to the points of a "smooth" line on one surface there corresponds a set of points on the other surface forming a smooth line. Let L_1 and L_2 be lines on S intersecting in the point p at which they include the angle α. Let the lines L_1' and L_2' be the images of the lines L_1 and L_2, respectively. Then L_1' and L_2' intersect in the point p', the image of p, at which they include the angle α'; we consider α' as the image of α and α as the original of α'.

(In the particular case of geographical maps, S is a part of the earth's surface and S' the corresponding part of a plane. Let us think of important lines on the earth's surface—coast lines, rivers, boundary lines, roads, railway lines—which are represented by corresponding lines on the map.)

(2) Now we are in a position to give a clear definition. We call a mapping *perfect* if it satisfies two conditions:

(I) All *lines* are reduced on the same scale.

(II) All *angles* are preserved.

Let us restate these two conditions with more detail.

(I) To the mapping there belongs a definite "scale" or fixed numerical ratio (for instance, $1:1\,000\,000$) in the following sense: If L', a line on the surface S', is the image of L, a line on the surface S, then the length of L' has that fixed numerical ratio ($1:1\,000\,000$ in our example) to the length of L, independently of the shape, size, and location of the lines.

(II) Each angle α' on S' is equal to the angle α on S of which α' is the image.

(3) Let us visualize the foregoing definition, let us see more concretely the details involved.

(3a) It is stated that the scale of a carefully done geographical map is $1:1\,000\,000$. This means that such is the scale *approximately*—but can the scale be the *same throughout the map exactly?* And, if such is the case, can the map also preserve the angles? That is the question.

(3b) If it is geometrically possible to map the surface S onto the surface S' on any fixed scale, it is, obviously, also possible to map a surface *geometrically similar* to S onto the surface S' on the scale $1:1$, that is, without reduction or enlargement. For example, let us assume that the planet on which we are living is an exact sphere. If any part of the earth's surface could be perfectly mapped onto a flat sheet of paper on the scale $1:1\,000\,000$, the corresponding part of a sphere of which the diameter is one millionth of the earth's diameter would be so mapped onto the same sheet that corresponding lines, original and image, would be of the same length throughout, and also corresponding angles would be equal.

(3c) We can roll a sheet of paper into a cylindrical or conical shape and, conversely, we can unroll the curved lateral surface of a cylinder, or of a cone, into a flat sheet. Such unrolling generates a perfect mapping of the curved cylindrical, or conical, surface onto a plane (imagine coastlines and rivers traced on the paper); lengths and angles are obviously preserved.

Yet could we similarly unroll into a plane a piece of a spherical surface, preserving all lengths and angles? We strongly suspect that it is not possible, and this suspicion may be based on experience, on observations we made in peeling apples or potatoes.

(4) Now we may perceive the kernel of our problem. Is it possible to map, with points corresponding one-to-one, a piece S of a spherical surface onto a piece S' of a plane so that *all lengths and all angles are preserved?*

We assume that (contrary to our expectation) such a mapping *is* possible and draw consequences from this assumption in the following (5) and (6).

(5) *Lengths are preserved.* Let p and q be two different points of S (of the sphere) and L any line on S connecting p and q; let p', q', and L' be the images in S' (in the plane) of p, q, and L, respectively. By our assumption, L and L' are of the same length. If L happens to be the *shortest line on the sphere* connecting p and q, shorter than any other connecting line, then, as lengths are preserved, L' must be shorter than any other line connecting p' and q', and so the shortest connecting line, *in the plane.* We know (the reader should know) that the shortest lines in the plane are straight lines, and the shortest lines on the sphere are arcs of great circles. The result of our consideration is: arcs of great circles on the spherical surface S are mapped onto segments of straight lines in the plane region S'. Especially, the sides of a spherical triangle which are arcs of great circles are mapped onto the sides of an ordinary triangle which are segments of straight lines.

(6) *Angles are preserved,* and so each angle of the spherical triangle just mentioned should be *equal* to the corresponding angle of the ordinary triangle. Yet this is impossible, since the sum of the three angles of an ordinary triangle is 180° whereas, as the reader should know, the sum of the three angles of a spherical triangle is *greater* than 180°.

A perfect mapping of the sphere onto the plane is impossible.

(7) The problem we have just solved may become a gateway both to practical applications (mapmaking) and to a great theory (a chapter of differential geometry centered in the "theorema egregium" of Gauss and reaching forward to general relativity). Here are a few points not too far beyond the high school level and closely connected with what we have just discussed; check them.

(7a) A plane (ordinary, Euclidean) triangle and a spherical triangle are so related that each side of one has the same length as the corresponding side of the other. Show that in this situation [considered in (5) and (6)] *each* angle of the spherical triangle is greater than the corresponding angle in the plane triangle. [That the sum of the angles of the first is greater than the sum of the angles of the second was the decisive remark of (6).]

(7b) The two conditions stated under (2) are not unrelated: (I) involves (II), that is if (I) is satisfied, (II) must be satisfied too.

(7c) Yet (II) does not involve (I). There are many mappings of the sphere onto the plane preserving all the angles, but in which the ratio of the length of a curve on the sphere to the length of the image curve in the plane is by no means constant [it cannot be constant, by virtue of the theorem proved in (5) and (6)].

(7d) There are mappings of the sphere onto the plane preserving all the areas (but they do not preserve the angles).

(7e) There are mappings of the sphere onto the plane preserving shortest lines, that is, mapping arcs of great circles onto segments of straight lines (but they do not preserve the angles).

(8) Yet I think that making explicite the role of continuity in the foregoing discussions and supplementing precise details in this respect would be too much above the level of the high school.

14.12. *What should we teach?* You, as a teacher, are paid by the community to teach the youngsters in your class. Therefore, your task is to teach what is profitable for the community and for the youngsters in your class.

You think that this advice does not amount to much. It may amount to more than you think. Just keep your task constantly in mind in short range as in long-range planning, in mapping out the next class period or the curriculum. Imagine that there is in your class a nice and clever boy, not yet spoiled by the school and not yet afraid of you who at any moment may ask you, honestly and naively: "But teacher, what is that good for?" If you try to picture to yourself what is in the mind of that nice boy and plan your teaching so that you can answer that critical question—or that he is constantly amused and challenged and has no opportunity to ask that critical question—you may become a better teacher.

I admit that the teacher's task is beset with temptations. For instance, we may be tempted to teach what is easy to teach, what is "teachable." Yet should we teach everything that is teachable? Is the teachable always profitable?

A clever trainer may teach a seal to balance a ball on its nose. But does such skill help the seal to catch more fish?

14.13. *The genetic principle.* Planning the curriculum involves more than choosing the facts and theories to be taught; we must also foresee in what sequence and by what methods those facts and theories should be taught. In this respect the "genetic principle" offers an important suggestion.

(1) The genetic principle of teaching can be stated in various ways, for instance: In teaching a branch of science (or a theory, or a concept) we should let the human child retrace the great steps of the mental evolution of the human race. Of course, we should not let him repeat in detail the thousand and one errors of the past, just the great steps.

This principle does not lay down a hard and fast rule; on the contrary, it leaves us much freedom of choice. What steps are great and what errors are negligible is a matter of interpretation. The genetic principle is a guide to, not a substitute for, judgement.

Yet in order to emphasize just this point, there may be some advantage in restating the principle more cautiously (and more vaguely): Having understood how the human race has acquired the knowledge of certain facts or concepts, we are in a better position to judge how the human child should acquire such knowledge. [We came very close to this formulation in ex. 14.10(3).]

(2) The genetic principle is supported by a biological analogy. The development of the individual animal retraces the evolutionary history of the race to which the animal belongs. That is, the embryo of the animal, as it passes through the successive stages of its development from the fertilized ovum to its adult form, resembles at each stage an ancestor of its race, and the sequence of its stages of development mirrors the sequence of its ancestors. If for "development of the individual animal" we say "ontogeny" and for "evolutionary history of the animal species" we say "phylogeny," we arrive at the concise form which the

German biologist Ernest Haeckel has given to his "fundamental biogenetic law": "Ontogeny recapitulates Phylogeny."

Such analogy is, of course, just a source of interesting suggestions and not a proof for the genetic principle of teaching, which itself should not be regarded as an "established principle" but just as a source of interesting suggestions.

(3) Thus, the genetic principle may suggest the principle of consecutive phases which we have discussed in sect. 14.4(3) and 14.5(3). In fact, in the historical development of various branches of science (of theories, of concepts) we may distinguish three phases. In an initial *exploratory* phase the first suggestive, but often incomplete or erroneous, ideas emerge from contact with the experimental material. In the next phase of *formalization* the material is ordered, appropriate terminology introduced, the laws recognized. In the last phase of *assimilation* the laws are seen in a broader context, extended, and applied.

Yet only the reading of the original works of great authors can really convince us of the genetic principle of teaching. Such reading may be like a brisk walk in the fresh air after the stale atmosphere of the textbooks. As James Clerk Maxwell wrote in the preface of his great *Treatise on Electricity and Magnetism:* "It is of great advantage to the student of any subject to read the original memoirs on that subject, for science is always most completely assimilated when it is in the nascent state."

(4) According to the genetic principle, the learner should retrace the path followed by the original discoverers. According to the principle of active learning, the learner should discover by himself as much as possible. A combination of the two principles suggests that the learner should *rediscover* what he has to learn—we have caught here a first glimpse of an important aspect of the educational process about which the reader should consult the two books of A. Wittenberg quoted in the bibliography.

14.14. *Lip service.* "General culture" is a catchword and, as a catchword, it is exposed to misuse. It is easy to pay lip service to "general culture." The most atrocious things can be done in the schools under the pretence that they are done for "general culture."

"General culture," "teaching to think," and "teaching problem solving" are catchwords which can be misinterpreted and misused inspite of the good substance behind them. Yet there is a difference in favor of "problem solving."

"Problem solving" can be explained not only in other general terms (equally liable to misinterpretation) but also by suggestive concrete examples (this book and my related writings attempt to present many such examples).

Moreover, lip service paid to problem solving can be more easily unmasked: "So, you are teaching problem solving—very interesting. What problems did you present to your class? What desirable mental attitudes do you expect to develop by such problems?"

14.15. *Confusion of levels.* "Contemporary mathematicians work much more with sets, operations, groups, fields, etc. than with oldfashioned geometry and algebra. Therefore, we must teach sets, operations, groups, and fields before those oldfashioned subjects."

This is an opinion. Here is a similar one:

"Contemporary American adults do many more miles driving a car than walking. Therefore, we must teach a baby to drive a car before he can walk."

14.16. *Isadora Duncan* was a celebrated dancer—as celebrated when I was young as Marilyn Monroe was in more recent years.

What is the connection between the lady and our subject? You see, it may seem such a good idea to let a team plan the curriculum and write the textbooks. The team consists of a university professor and a high school teacher. We may expect a splendid result in which the professor's mathematical perspective is coupled with the teacher's experience in handling high school classes. Yes, yes, but. . . .

When I was young everybody knew a certain story about Isadora Duncan. Allegedly, she offered something like marriage to Bernard Shaw: ". . . and think of the child who would have your brains and my looks." "Yes, yes," said Bernard Shaw "but what a calamity if the child had my looks and your brains."

Perhaps you too have seen some books recently which mirror the teacher's mathematical perspective coupled with the professor's experience in handling high school classes.

14.17. *Levels of knowledge.* In his "Treatise on the Improvement of the Mind" (Tractatus de Intellectus Emendatione), the philosopher Benedict Spinoza distinguishes four different levels of knowledge.[18] Spinoza exemplifies the four levels by four different ways of understanding the Rule of Three. In the following, (1) to (4), "rule" stands for any mathematical rule the reader learned some time ago and, if possible, for one that he has learned by stages, understanding it better at each stage.

(1) A student has learned a rule by heart, accepting it on authority, without proof, but he is able to use the rule, can apply it correctly. Then we say that he has *mechanical knowledge* of the rule.

(2) The student tried the rule on simple cases and convinced himself that it works correctly in all cases he has tested. He thus has *inductive knowledge* of the rule.

(3) The student has understood a demonstration of the rule. He has *rational knowledge* of the rule.

(4) The student conceives the rule clearly and distinctly and is so convinced of it that he cannot doubt that the rule is true. He has *intuitive knowledge* of the rule.

(5) I do not know whether the passage of Spinoza paraphrased in the foregoing has been noticed in the pedagogical literature. At any rate, the distinction between various levels of knowledge should be well understood by the teacher. Thus the curriculum requires of him that he should teach such and such chapter of mathematics in such and such grade. Yet what level of knowledge should the students attain? Is mechanical knowledge sufficient? Or should the teacher attempt to lead his students to intuitive knowledge? We have here two different aims, and it makes a great difference both for the teacher and for the students which one is adopted.

---

[18] See e.g. *Philosophy of Benedict Spinoza,* translated by R. H. H. Elwes, p. 7. The following is a very free paraphrase of Spinoza's text; especially, the names for the levels are added.

(6) When we consider the various levels of knowledge distinguished by Spinoza from the teacher's standpoint, we are led to several questions. How could we bring the students to this or that level? How could we test whether the students have attained this or that level? These questions are most difficult to answer in regard to the intuitive level.

(7) We could, of course, distinguish more than four levels of knowledge, and there is one level whose consideration may deserve the attention of teachers (and of learners, especially of ambitious learners who aspire to become scientists)— well anchored, well connected, well cemented, in one word *well-organized knowledge*.[19]

The teacher who aims at well-organized knowledge should be careful, in the first place, in introducing new facts. A new fact should not appear out of nowhere, it should be motivated by, referred to, connected with the world around us and the existing knowledge, the daily experience, the natural curiosity of the student.

Moreover, when the new fact has been understood, it should be used to solve new problems, to solve old problems more simply, to shed light on things already known, to open new perspectives.

The ambitious learner should carefully study a new fact; he should turn it over and over, consider it under various aspects, scrutinize it from all sides, and try to fit it into his existing knowledge at the best place where it is most conveniently connected with related facts. Then he will be able to see that new piece of knowledge with the least effort, the most intuitively. Moreover, he should try to expand and enlarge any newly acquired knowledge by application, generalization, specialization, analogy, and in all other ways.

(8) As dedicated teachers we may find ways to anchor a new fact in the experience of the student, connect it with formerly learned facts, cement its knowledge by applications. We can only hope that the student's well-anchored, well-connected, well-cemented, well organized knowledge will finally become intuitive.

14.18. *Repetition and contrast.* If you like both music and teaching, you may observe various resemblances between them, and your observations, even if they are not scientifically perfect, may improve your teaching. They may lead you to presenting the material you have to offer in a more artistic and more effective arrangement.

Why is that so? Repetition and contrast play a role in all arts, also in the teacher's art, but their role is most conspicuous in music. Hence the foreshadowing, the development, the repetition, alternation, and variation of themes in a musical composition may well suggest the analoguous treatment of themes in the classroom or in a literary composition.

14.19. *Inside help, outside help.* In planning and writing this book I had in mind the task of the high school mathematics teacher and especially the following situation. The teacher proposes a problem to his class; the students should learn from their own work, and the problem should be solved in class discussion. This situation requires careful handling. If the teacher helps too little, there will be

---

[19] *Cf.* ex. 12.3, and also the preface and the whole plan of G. Pólya and G. Szegö, *Aufgaben und Lehrsätze aus der Analysis.*

no progress. If the teacher helps too much, there will be no opportunity for the students to learn from their own work. How should the teacher avoid the horns of this dilemma? How much should the teacher help the students?

There is a better question; we should not ask "how much," we should ask "how." *How* should the teacher help the students? There are various ways to help.

(1) There are cases in which the teacher must ask several questions and repeat his questions several times until he succeeds in extracting a little work from the students. In the following dialogue a dotted line . . . . . . indicates the silence of the students. The discussion has been going on for some time as the teacher says:

"Tell me again: What is the unknown?"
'The length of the line AB.'
"How can you find this kind of unknown?"

. . . .
"How can you find the length of a line?"

. . . .
"From which data could you derive the length of a line?"

. . . .
"Did not we solve such problems before? I mean problems in which the unknown was the length of a line?"
'I think we did'.
"How did we proceed in such a case? From which data did we compute the unknown length?"

. . . .
"Look at the figure. You see the line AB, don't you? The length AB is unknown. Of which lines is the length known?"
'AC is given.'
"Good! Is there any other given line?"
'BC is also given.'
"Look at the lines AB, AC, and BC—look at their *configuration*. How would you describe this configuration?"
'AB, AC, and BC are the sides of $\triangle ABC$.'
"What kind of triangle is $\triangle ABC$?"

. . . .

Well, there are cases in which a teacher must be infinitely patient.

(2) A less patient teacher could proceed quite differently and tell the students right away: "Apply the theorem of Pythagoras to the right triangle $\triangle ABC$."

(3) What is the difference between the two procedures (1) and (2)?

The most obvious difference is that (1) is long and (2) is short.

Yet we should also observe another difference: (1) offers more opportunity to the student than (2) to contribute to the solution something of his own.

Yet there is a more subtle difference.

The questions and suggestions offered by the teacher in procedure (1) could have *occurred to the student himself*. If you look at them closely, you may notice that several of these questions and suggestions are *tools* which the problem solver can use also for other problems, in fact, for wide classes of problems. These tools are at the disposal of everybody—of course, the more experienced, the "methodologically better prepared" problem solver handles them more deliberately and more skillfully.

Yet the action advised by the teacher in procedure (2) is not a tool a priori at the disposal of the problem solver—it is the solution itself, or it is almost that. It is a specific action suggested without reference to any general idea.

Let us call *inside help* such help as the problem solver earnestly concerned with his problem and familiar with methodological ideas has a good chance to find by himself. Let us call *outside help* such help as has little relation to methodological ideas—the problem solver has little chance to elicit such help by methodical work. I think that the most important difference between the procedures (1) and (2) is that the teacher offers inside help in the former and only outside help in the latter.

(4) If we accept the principle of active learning we must prefer inside help to outside help. In fact, the teacher should give outside help only in the last resort, when he has exhausted all obvious inside suggestions without result, or when he is pressed for time.

Outside help has very little chance to be instructive—appearing out of the blue, as *deus ex machina,* it can easily be disappointing.[20] Inside help may be the most instructive thing the teacher can offer; the student may catch on, he can realize that the question helps and that he could have put that question to himself by himself. And so the student may learn to use that question; the voice of the teacher may become for him an inner voice which warns him when a similar situation arises.[21]

To give inside help, the teacher may use all the "stereotyped" questions and recommendations collected in chapter 12—for this reason, chapter 12 may become for him the central chapter of the book. Of course, he must first become intimately acquainted with the situations to which those questions and recommendations are applicable. This book has been planned and written to help the teacher in this task.

14.20. In my class, when I get the impression that I have spoken too long without interruption and I should now ask a question of the audience, I am apt to remember a German jingle of which here is an approximate translation:

All are sleeping just one is preaching:
Such performance is called here "teaching."

14.21. *How difficult is it?* Both the scientist and the teacher may be led to this question, the one when he is struggling with a problem, the other when he is about to propose it to his class. In answering the question we must rely much more on "feeling" than on any clear argument. Still, now and then we arrive at assessing the degree of difficulty of a problem quite well. The scientist's feeling may be justified by the outcome of his research and the teacher's feeling by the outcome of an examination.

In most circumstances, clear arguments can contribute but little to the evalua-

---

[20] *Cf.* the end of sect. 3.2.
[21] *Cf.* HSI *passim,* especially sect. 17, Good questions and bad questions, pp. 22–23.

tion of the degree of difficulty of a problem, yet even that little deserves careful consideration.

(1) *Size of the region of search.* An offense has been committed (for instance, one of the children has broken a window) and we know that the offender is one of a set of n persons. Other things being equal, the difficulty to find the offender obviously increases with n. Generally speaking, we may expect that the difficulty of a problem increases with the size of the adequate region of search. (*Cf.* sect. 11.6.)

(2) *Number of items to be combined.* The students have to solve a problem which needs the application of n different rules introduced in the last chapter with which the students are much less familiar than with the foregoing chapters of the course. Under such circumstances, other things being equal, the difficulty of the problem obviously increases with n. Generally speaking, we may expect that the difficulty of a problem increases with the number of items which we have not combined before but which we have to combine to solve the problem.

(3) The foregoing considerations may help us to judge the difficulty of a problem a priori, before trying it. Judging the difficulty a posteriori, after having tried to solve the problem, involves, more or less explicitely, ideas of *statistics;* here is a schematic example. In an examination, of two problems proposed to 100 students, one was solved by 82 students, the other by 39 students. Obviously, the latter was more difficult *for this group* of 100 students. Will it also be more difficult for the next group of students? Yes, says the statistician, this has to be expected with such and such a degree of confidence provided that there are *no nonrandom differences* between the two groups. And there is the rub. In educational matters there are too many scarcely controllable circumstances which have a great influence so that the distinction between "random" and "nonrandom" becomes highly elusive. Thus the presentation of a certain point in the course, the insistence of the teacher on that point, the mood of the teacher, and many other unforseeable circumstances may incontrollably influence the outcome of an examination—they may influence it much more than that component of the situation about which we have desired to obtain information by statistical experiment. We have touched here but lightly on one of the many reasons which should make us cautious or even suspicious when we are dealing with educational statistics.

Of course, when the mathematician deals with a problem that was proposed two hundred or two thousand years ago yet still remains unsolved, he has a rather good "statistical" reason to suspect that the problem is hard. (The theory of numbers abounds in such problems.)

14.22. *Difficulty and educational value.* It is difficult to assess the difficulty, and still more difficult to assess the educational value, of a problem. Yet the teacher must try to assess both when he is about to propose the problem to his class.

The teacher can be assisted in his task by a classification of high school level problems. The credit for a classification of this kind is due to Franz Denk.[22] The following classification is somewhat different; it distinguishes four types of problems.

[22] Franz Denk, Werner Hartkopf, and George Pólya, pp. 39–42; see the Bibliography.

(1) *One rule under your nose.* The problem can be solved by straightforward mechanical application of a rule or by straightforward mechanical imitation of an example. Moreover, the rule to apply or the example to follow is thrust under the nose of the student; typically, the teacher proposes such problems at the end of the hour in which he has presented the rule or the procedure. A problem of this type offers practice but nothing else; it may teach the student to use that particular rule or procedure, but has little chance to teach him anything else. And there is the danger that, even of that single rule, the student will acquire just "mechanical," and not "insightful," knowledge.

(2) *Application with some choice.* The problem can still be solved by the application of a rule learned in class or the imitation of an example shown in class, yet it is not so immediately obvious which rule or example should be used; the student needs some mastery of the material covered in the last weeks and some judgement to find the usable item in a certain limited region of search.

(3) *Choice of a combination.* To solve the problem, the student must combine two or more rules or examples shown in class. The problem need not be too difficult if a somewhat similar (but not the same!) combination has been discussed in class. Of course, if the combination is quite novel, or if many pieces of knowledge must be combined, or pieces of knowledge from chapters wide apart, the problem may demand a higher degree of independence and may become quite difficult.

(4) *Approaching research level.* It is scarcely possible to draw a sharp line of demarcation between the kind of problem we have just considered [under (3)] and research problems. The examples and discussions of chapter 15 attempt to outline some characteristics desirable in "research problems on the classroom level."

On the whole, as the degree of difficulty increases along the lines presented in ex. 14.21(1) and (2), also the educational value of the problem has a chance to increase, especially if our educational aim is "teaching to think," and we judge values from the standpoint of this aim.

14.23. *Some types of problems.* For the use of teachers who wish to interrupt now and then the monotonous sequence of routine problems dished out by the textbooks, I collected elsewhere some types of nonroutine problems (MPR, v. 2, p. 160). I wish to add here one more problem of the "You may guess wrong" type (ex. 14.25) and one new type, the "red herring" problem. This latter type diverts the attention from the main point, from the most natural or most efficient procedure, by some conspicuous but irrelevant feature. "Red herring" problems should be sparingly used; they should be proposed only to students clever enough to see the joke and to learn better the relevant point by having brushed aside irrelevancies. See ex. 14.24.

14.24. Find the remainder of the division of the polynomial

$$x^3 + x^5 + x^7 + x^{11} + x^{13} + x^{17} + x^{19}$$

by the polynomial $x^2 - 1$.

14.25. Two spheres are tangent to each other. They are separated from each other by their *principal* common tangent plane that passes through their point of contact. They have infinitely many other common tangent planes which envelop their common *tangent cone*. This cone touches each sphere along a circle, and the part of the cone between these two circles is the lateral surface of a *frustum*.

Being given t, the slant height of the frustum, compute:

(1) the lateral area of the frustum

(2) the area of that piece of the "principal" tangent plane that lies within the tangent cone.

(Are there enough data to determine the unknowns?)

14.26. *A term paper.* For terminology and notation, see "Mathematics and Plausible Reasoning," vol. 1, pp. 137–138, ex. 33; you may also be able to use appropriate points from ex. 34–54 on pp. 138–141 and their solutions on pp. 249–256.

Consider

(a) a right prism,

(b) a right pyramid, and

(c) a right double pyramid.

The base of each of these three solids is a regular polygon with n sides, and each solid is so circumscribed about a sphere that the point of contact on each face of the solid is the center of gravity of that face.

(1) Find the ratio of the area of the base B to the area of the total surface S of the solid for (a), (b), and (c).

(2) Compute S^3/V^2 for (a), (b), and (c); V stands for volume.

(3) Collect in a table the numerical values of the ratios computed under (2) for $n = 3, 4, 5$, and 6.

(4) Describe the limiting case $n \to \infty$ for (a), (b), and (c). Find the limit of the ratios computed in (2) and add their numerical values to the table (3).

(5) Consider the (unsolved) problem: "Find the polyhedron with a given number F of faces and a given volume that has the minimum surface area."

Formulate a "plausible" conjecture for $F = 4, 6, 8, 12$, and 20; explain why it is plausible, and find indications for or against it in your foregoing work.

(6) Choose carefully a particular problem from the foregoing work that could be appropriately treated in a solid geometry class in the high school.

Formulate it clearly.

Note such "inside" questions and suggestions (from the list "How to Solve It," HSI, pp. XVI–XVII, for instance) [*cf.* chapter 12] as have a good chance to be efficiently used for the problem chosen.

Represent the progress of the solution of the chosen problem diagrammatically (as we did it in class for the volume of the frustum.) [*Cf.* chapter 7.]

(7) How would you "sell" (justify the choice of, explain the general interest of) the subject of this term paper to a class of bright students in a high school, or in a teacher-training institution? (Name the level, and be efficient but concise—please!)

(8) A *diagonal* of a convex polyhedron is a straight-line segment joining two vertices that, except for its two endpoints, lies completely inside the polyhedron (not on the surface). Let D denote the number of diagonals of the polyhedron.

Compute D

(a) for each of the five regular solids,

(b) for a polyhedron all F faces of which are triangles,

(c) for a general polyhedron, being given that it has F_n polygonal faces with n sides, $n = 3, 4, 5, \ldots,$

$$F_3 + F_4 + F_5 + \cdots = F$$

(9) (Optional) If the foregoing has suggested to you some mathematical idea which seems to you relevant, even if incomplete for the moment, note it here—but *very* clearly and concisely.

(The foregoing is a typical example of a "take-home final" such as I am accustomed to propose to my classes offered to high school teachers. For (5) see below ex. 15.35, for (8) ex. 15.13. To answer (6) and (7), some participants presented a dialogue between the teacher and the students similar to some dialogues in this book and especially in HSI. Some of these dialogues were very well done. The references in square brackets were added in print.)

14.27. *On talks at mathematical meetings. Zermelo's rules.* The situation of the speaker in a mathematical meeting resembles a little, and differs a great deal from, the situation of the teacher in the classroom. The speaker, as the teacher, intends to impart some information, but his audience consists of his peers and possibly his superiors, not of students. The situation of the speaker is not easy, also his performance is not often successful. This is not so much his fault, rather the fault of the vast expanse of mathematics. Any one mathematician can master only a small parcel of present day mathematics, and usually knows very little about the other small parcel that the next mathematician has mastered.

(1) Ernest Zermelo, whose name will be forever linked to the important "axiom of choice" of general set theory, liked to spend some of his time in coffee-houses. His conversation there was interspersed with sarcastic remarks about his colleagues. Commenting on an address which had great success at a recent mathematical meeting, he criticized the speaker's style and eventually condensed his disapproval into two rules according to which, he mockingly asserted, that address must have been constructed:

I. *You cannot overestimate the stupidity of your audience.*

II. *Insist on the obvious and glide nimbly over the essential.*[23]

Zermelo's personal remarks were often witty; very unjust on the whole, but striking and revealing about some particular point. So was the criticism implied by the two rules; I had to laugh and I could not forget the rules. Years later I realized that these rules, suitably interpreted, give often applicable sound advice.

(2) The speaker in a mathematical meeting usually treats his audience as if everybody knew everything about the subject he is discussing—especially each detail of his latest paper. Of course, just the contrary is the case, and the speaker should realize it. It would be better for him to overestimate than underestimate the lack of knowledge of the audience about the subject. In fact, the speaker

---

[23] In the original German: *I. Du kannst Deine Hörer nicht dumm genug einschätzen. II. Bestehe auf dem Selbstverständlichen und husche über das Wesentliche hinweg.*

could greatly profit from an interpretation of Zermelo's first rule: "You cannot overestimate the lack of knowledge of your audience about the things you are talking about."

(3) What is essential in the mathematician's work?   Each detail of the proof is essential, but it is hardly possible to present adequately all details of a long and difficult proof to a mathematical meeting.   Even if the speaker succeeded in hurrying through all the details, nobody would be able to follow.   Therefore, "glide nimbly over the essential," over the details of the proof.

Yet even a long proof may hinge on a central remark which in itself is intuitive and simple.   A good speaker should be able to detach from the proof some decisive remark and make it intuitive and obvious until everybody in the audience can understand it, take it home, and keep it for possible later use.   In so doing the speaker succeeds in imparting useful information and, in fact, follows Zermelo's second rule: "Insist on the obvious and glide nimbly over the essential."

14.28. *Epilogue.*   When I was young I devoured the novels of Anatole France. More than by the stories themselves I was attracted by the tone in which they are told, the voice of a sage who looks at all things human with a delicate irony mingled with pity.

Anatole France has a word about a subject we have been discussing: "Do not try to satisfy your vanity by teaching them great many things.   Awake their curiosity.   It is enough to open the minds, do not overload them.   Put there just a spark.   If there is some good inflammable stuff it will catch fire."   (Le jardin d'Epicure, p. 200.)

There is a great temptation to paraphrase this passage: "Do not try to satisfy your vanity by teaching high school kids a lot of . . . just because you wish to make people believe that you understand it yourself. . . ."   Yet let us resist temptation.

CHAPTER 15

GUESSING AND SCIENTIFIC METHOD [1]

Non-mathematical induction plays an essential role in mathematical research.

I. SCHUR: *Inaugural Dissertation,* Berlin 1901, Thesis I.

It will always be difficult, in any branch of knowledge, to describe with some approximation to the truth the method followed by the inventors. . . . Nevertheless, concerning the mental procedures of mathematicians there is a simple remark extensively confirmed by the history of science: observation has an important place and plays a great role in their procedures.

CHARLES HERMITE: *Œuvres,* vol. IV, p. 586.

Observation is the abundant source of invention in the world of subjective realities just so as it is in the world of phenomena perceptible by the senses.

CHARLES HERMITE: *Hermite et Stieltjes, Correspondance,* vol. I, p. 332.

15.1. Research problems on the classroom level

The teaching of mathematics should acquaint the students with all aspects of mathematical activity as far as possible. Especially it should give opportunity to the students for independent creative work—as far as possible.

The activity of the expert mathematician, however, differs a great deal and in several respects from the usual classroom activity. We shall be in a better position to see which differences deserve particular em-

---

[1] This chapter is dedicated to my friend and colleague, Charles Loewner.

phasis after a few examples. The following examples show that a good teacher can offer something approaching the experience of independent inquiry even to an average class by choosing appropriate problems and presenting them in the appropriate way.

15.2. Example

"Being given L, the length of the perimeter of an isosceles right triangle, compute its area A." This is the usual kind of problem printed in the usual textbook. It is not even a bad problem, just not too interesting when it is presented by itself, isolated from analogous problems. Compare it with the following presentation, and see the difference.

"In those legendary pioneer days" said the teacher "when land was plentiful and almost everything else very scarce, a man in the Middle West had hundreds of acres of flat grassland, but only one hundred yards of barbed wire. He intended to use the whole length of this wire to fence in a piece of his land. He thought of various shapes and wondered how many square yards would be the fenced in area."

"Well, which shape would you prefer? But remember, you have to compute the area, and so you rather choose some easy shape."

– A square.

– A rectangle with sides of 20 and 30 yards.

– An equilateral triangle.

– An isosceles right triangle.

– A circle.

"Very good. May I add a few shapes:

A rectangle with sides of 10 and 40 yards.

An isosceles triangle with sides of 42, 29, and 29 yards.

An isosceles trapezoid with sides of 42, 13, 32, and 13 yards.

A regular hexagon.

A semicircle."

"All these figures are *isoperimetric,* that is, of equal perimeter; the perimeter of each is supposed to be 100 yards. Compute the areas in square yards, and arrange the ten figures according to their areas, beginning with the largest and ending with the smallest. By the way, before computing you may try to guess which area will be the largest and which one will be the smallest."

This problem may be proposed as homework to an average high school class at an appropriate stage of the curriculum. The solution consists of the following list:

Circle	795	Rectangle 30, 20	600
Regular hexagon	722	Semicircle	594
Square	625	Equilateral triangle	481

Trapezoid 42, 13, 32, 13   444   Triangle 42, 29, 29   420

Isosceles right triangle   430   Rectangle 40, 10   400

"Are there any questions?"

15.3. Discussion

The point of our problem is to bring to the attention of the students the list of figures and areas which constitutes its solution; the observation of this list should suggest various remarks to the students. The more spontaneously these remarks emerge, the better it is. Yet, if they are slow in coming, the teacher may contribute to the discussion by well-placed, gently prodding questions, such as the following.

"Have you any comment on the list?"

"The circle leads the list. Have you any comment on this?"

"There are several triangles in the list, also several quadrilaterals. Which figure leads the list of quadrilaterals? What about the triangles?"

"Yes, it may be so as you say, but have you proved it?"

"If you have not proved it what reason have you to believe it?"

"A triangle may be considered as a degenerate quadrilateral, with one vanishing side (or with one angle of 180°). Does this remark contribute to your argument?"

Eventually the students should arrive, sooner or later, by their own means as far as possible, to remarks in the direction of the following.

The list suggests that of all plane figures of equal perimeter the circle has the maximum area.

The list suggests that of all quadrilaterals of equal perimeter the square has the maximum area.

The list suggests that of all triangles of equal perimeter the equilateral triangle has the largest area.

The list suggests that of all polygons with a given number n of sides and with a perimeter of given length the regular polygon has the largest area.

Another suggestion of the list: If two regular polygons have the same perimeter, the one with more sides has the larger area. (The more a polygon resembles the circle the larger seems to be its area.)

None of these statements is proved by the list, which yields, however, some kind of reason to believe in them, more or less.

Our experience may also suggest more general insights, such as the following: Verification in more cases gives us a stronger reason to believe.

There may be other relevant remarks, and some of the foregoing remarks may come more readily after more examples.

15.4. Another example

"The Greeks knew" said the teacher "a remarkable proposition on the area of the triangle which we call today 'Heron's formula' and express by the equation

$$A^2 = s(s - a)(s - b)(s - c)$$

where A stands for the area of the triangle, a, b, and c for the lengths of the three sides, and

$$s = \frac{a + b + c}{2}$$

for the semiperimeter.

"The proof for Heron's formula is not quite simple, and I do not wish to enter into it today. In absence of a proof, however, we cannot be certain that the equation as it stands is correct—my memory may have failed me as I wrote it down. Could you check the formula? How could you check it?"

– I would try it on the equilateral triangle.

In this case $a = b = c$, $s = 3a/2$, and the formula yields the correct result. "What else could we do?"

– I would try it on a right triangle.

– I would try it on an isosceles triangle.

In the first case $a^2 = b^2 + c^2$, in the second case $b = c$, and in both cases the formula yields, after some algebraic manipulation, the correct result. (The reader should work out the details.)

"Do you like it?"

– Yes, it checks.

"Could you think of a further particular case that we should check?"

"What about the degenerate triangle? I mean the extreme, or limiting, case in which the triangle collapses into a straight line segment."

In this case $s = a$ (or b, or c) and the formula obviously yields the correct result.

– Please, teacher, in how many cases must we check a formula, if we want to be sure that it is right?

The reader may picture to himself the discussion started by the last question.

15.5. Graphic representation of the inductive argument's progress

What is, and what is not, accomplished by the successive verifications of the proposed formula in sect. 15.4? Each verification deals with a certain triangular shape, and so a survey of all such shapes may contribute to elucidating our question.

Let x, y, and z denote the sides of a variable triangle listed in order of increasing length, so that

$$0 < x \leqq y \leqq z$$

Then we must have

$$x + y > z$$

Now, as only the shape of the triangle matters, and not its size, we may assume that

$$z = 1$$

Thus we have three inequalities

(1) $$x \leqq y, \quad y \leqq 1, \quad x + y > 1$$

Let us now represent the triangle with sides x, y, and 1, or triangle $(x, y, 1)$ for short, by the point (x, y) in a plane the rectangular coordinates in which are x and y. Each one of the three inequalities (1) restricts the point (x, y) to a half-plane (including the boundary line in the first two cases, excluding it in the third case). Jointly the three inequalities (1) characterize a set of points—the common part or *intersection* of the three half-planes. This intersection is a triangle, see Fig. 15.1, with vertices (1, 1), (0, 1), and $(\frac{1}{2}, \frac{1}{2})$ [including the vertex (1, 1) and the two adjacent sides, but excluding the two other vertices and the third side]. This triangular area represents the totality of triangular shapes; the point (x, y) represents the triangle $(x, y, 1)$ and different points represent different shapes.

How are the particular cases considered in sect. 15.4 located in Fig. 15.1?

First, we have verified the proposed formula for the equilateral triangle. Such a triangle is (1, 1, 1)—we mark with this symbol the corresponding point (1, 1) in Fig. 15.2.

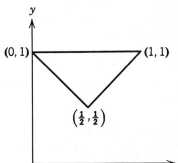

Fig. 15.1. The totality of triangular shapes.

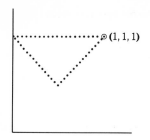

Fig. 15.2. Verified for the equilateral triangle.

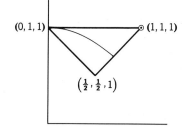

Fig. 15.3. . . . and for right triangles.

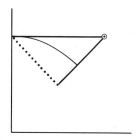

Fig. 15.4. . . . and for isosceles triangles.

Fig. 15.5. . . . and for degenerate triangles.

Then we verified the formula for right triangles. If $(x, y, 1)$ is a right triangle, its greatest side 1 must be its hypotenuse and, therefore,

$$x^2 + y^2 = 1$$

Hence the right triangles are represented in Fig. 15.3 by a circular arc (of the unit circle).

Then we considered isosceles triangles. We should here distinguish between two kinds of isosceles triangles: either the two longer sides are equal and so

$$y = 1$$

or the two shorter sides are equal and so

$$x = y$$

Hence, the points representing isosceles triangles fill two boundary lines in Fig. 15.4 (fully drawn there—they were just dotted in the foregoing Fig. 15.2 and 15.3).

Finally, for a degenerate (collapsing) triangle $(x, y, 1)$

$$x + y = 1$$

Such "triangles" are represented by the third boundary line, fully drawn in Fig. 15.5 (but just dotted in the foregoing Figs. 15.2, 15.3, and 15.4).

Surveying the sequence of diagrams from Fig. 15.2 to Fig. 15.5, we may visualize the progress of the inductive argument. At the beginning, in Fig. 15.2, one point was enough to represent the extent of verification. Then more and more full lines appear in the diagram, indicating more and more classes of cases in which the verification has succeeded.

The points representing the triangular shapes for which the formula in question has been verified, are distributed along lines. Yet the formula remains unverified for the "bulk" of the triangular shapes represented by the area not covered by those lines. Still, since the formula has been verified along the whole boundary and also along a crossing line, we may reasonably expect that it will turn out correct in all cases. The part suggests the whole and suggests it strongly.

15.6. A historic example

We are going to investigate a problem of solid geometry by following in the footsteps of two great mathematicians. I shall tell their names later, but not too early, otherwise a good point of my story could be spoiled.

(1) *Analogy suggests a question.* A polyhedron is enclosed by plane faces just so as a polygon is enclosed by straight sides; polyhedra in space are analogous to polygons in a plane. Yet polygons are simpler, more accessible than polyhedra, a question about polygons has a good chance to be much easier than the corresponding question about polyhedra. When we know a fact about polygons, we should try to discover an analogous fact about polyhedra; in doing so we have a good chance to hit upon a stimulating question.

For instance, we know that the sum of the angles in a triangle is the same for all triangles: it is, independently of the size and shape of the triangle, equal to 180°, or two right angles, or π (in radians; we shall prefer this last way of measuring angles). More generally, the sum of angles in a polygon with n sides is $(n - 2)\pi$. Now, let us try to discover an analogous fact about polyhedra.

(2) *We try to exhaust the possibilities.* Our goal, however, is not quite clear. We wish to find out something about the sum of the angles in a polyhedron—but what angles?

Each edge of the polyhedron is associated with a *dihedral* angle, included by the two faces adjacent to that edge. Each vertex of the polyhedron is associated with a *solid* angle included by all the faces (three or more) adjacent to that vertex. Which kind of angle should we consider? Has the sum of all angles of the same kind some simple property? What about the sum of the six dihedral angles in a tetrahedron? What about the sum of the four solid angles in a tetrahedron?

It turns out that none of the two last named sums is independent of the shape of the tetrahedron (see ex. 15.14). How disappointing! We expected the tetrahedron to behave like a triangle.

Still, we may be able to save our original idea. We have not yet exhausted all possibilities. There are, in a polyhedron, angles of still another kind (this kind is the most familiar, in fact): each face enclosed by n sides (by n edges of the polyhedron) contains n interior angles. Let us call such angles *face angles,* and let us try to find the *sum of all face angles* of the polyhedron. Let $\Sigma\alpha$ denote the desired sum; see Fig. 15.6.

(3) *We observe.* If we see no other approach to our problem, we can always attack it experimentally: we take a few polyhedra and we compute $\Sigma\alpha$ (the sum of the face angles) for each. We may begin with the cube, see Fig. 15.7a. Each face of the cube is a square; the sum of the four angles in a square is 2π. As there are six faces, $\Sigma\alpha$ is

$$6 \times 2\pi = 12\pi$$

for the cube. We can handle just as easily the tetrahedron and the octahedron, see Fig. 15.7b and c.

The three polyhedra we have examined so far are regular. Let us examine, for a change, some nonregular polyhedron, for instance a 5-prism (a prism with pentagonal base, see Fig. 15.7d). This prism has two kinds of faces: there are five parallelograms and two pentagons. Therefore, $\Sigma\alpha$ is

$$5 \times 2\pi + 2 \times 3\pi = 16\pi$$

Fig. 15.6. Face angle.

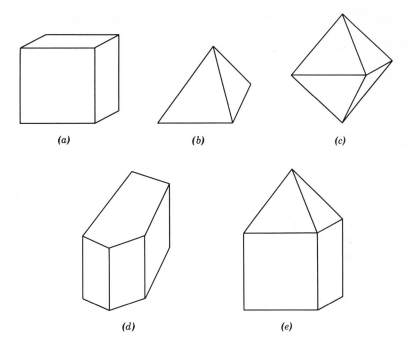

(a)　　　　　*(b)*　　　　　*(c)*

(d)　　　　　*(e)*

Fig. 15.7. Polyhedra.

for the 5-prism. Then let us take a polyhedron which we see less often in a classroom, see Fig. 15.7e: a pyramid is placed on a cube as "roof"; the resulting "tower" has 9 faces, 5 squares, and 4 triangles; hence its $\Sigma\alpha$ equals

$$5 \times 2\pi + 4 \times \pi = 14\pi$$

We collect our observations in Table I; to render more recognizable the polyhedra considered, let us note for each the number of its faces, F.

TABLE I

Polyhedron	F	$\Sigma\alpha$
Cube	6	12π
Tetrahedron	4	4π
Octahedron	8	8π
5-Prism	7	16π
Tower	9	14π

Do you notice something worth noticing—some law or pattern or regularity?

(4) *We observe, guided by an idea.* It is not surprising that, at this stage of the game, we see nothing striking in the observational material hitherto collected; observation without some leading idea seldom produces noteworthy results.

In reflecting on our own procedure, we may find our way out of this impasse. In (3), we have repeatedly computed the grand total of the face angles $\Sigma\alpha$ by taking first the sum of those angles that belong to the same face—we know this sum precisely, in fact the knowledge of this sum was the starting point of our inquiry. Let us now take, for a change, the sum of those angles that have the same corner of the polygon as vertex. We do not know this sum precisely, but we know that it is less than 2π, a full plane angle. (We now restrict ourselves explicitly to *convex* polyhedra; the fact quoted is intuitive, but you may look up Euclid XI 21 for a proof.) Let V denote the number of vertices of the polyhedron considered; we perceive that the grand total of the face angles

$$\Sigma\alpha < 2\pi V$$

Let us verify this relation on our collected material! We extend our Table I into Table II.

TABLE II

Polyhedron	F	$\Sigma\alpha$	V	$2\pi V$
Cube	6	12π	8	16π
Tetrahedron	4	4π	4	8π
Octahedron	8	8π	6	12π
5-Prism	7	16π	10	20π
Tower	9	14π	9	18π

In fact, throughout Table II, $2\pi V$ is larger than $\Sigma\alpha$, and we can hardly fail to observe that the difference is constant

$$2\pi V - \Sigma\alpha = 4\pi$$

Is this a coincidence? A mere coincidence seems unlikely, and so we can hardly resist the temptation to guess that the relation observed holds not only in the few cases we have examined, but generally for all convex polyhedra. Thus we arrive at the *conjecture*

$$(?) \qquad \Sigma\alpha = 2\pi V - 4\pi$$

The query in parenthesis in front of the stated relation should remind us that it is not proved, merely conjectured.

(5) *We test our conjecture.* Our observation, guided by a lucky remark, produced a remarkable conjecture—but is it true?

Let us check a few more cases. There are two more regular polyhedra
to consider, the dodecahedron and the icosahedron, with $F = 12$ and
$F = 20$, respectively. Furthermore, we could consider a general prism,
the n-prism, the base of which is a polygon with n sides, then the
n-pyramid with a base of the same nature, and the n-double-pyramid;
the last consists of two n-pyramids standing on opposite sides of a common base (which is *not* a face of the resulting double pyramid). The
reader can easily extend Table II to the solids mentioned.

EXTENSION OF TABLE II

Polyhedron	F	$\Sigma\alpha$	V	$2\pi V$
Dodecahedron	12	36π	20	40π
Icosahedron	20	20π	12	24π
n-Prism	$n + 2$	$(4n - 4)\pi$	$2n$	$4n\pi$
n-Pyramid	$n + 1$	$(2n - 2)\pi$	$n + 1$	$(2n + 2)\pi$
n-Double-pyramid	$2n$	$2n\pi$	$n + 2$	$(2n + 4)\pi$

The conjectural relation (?) is verified in all cases examined, which
is gratifying but does not amount to a proof.

(6) *Reflections on the procedure followed.* In computing $\Sigma\alpha$ we have
used the same procedure several times: we started by taking the sum
of such angles as belong to the same face. Why not apply this procedure *generally*?

To follow up this reflection we introduce suitable notation. Let

$$s_1, s_2, s_3, \ldots, s_F$$

denote the number of sides of the first, the second, the third, . . . , the
last face, respectively. With this notation

$$\Sigma\alpha = \pi(s_1 - 2) + \pi(s_2 - 2) + \cdots + \pi(s_F - 2)$$
$$= \pi(s_1 + s_2 + \cdots + s_F - 2F)$$

Now, $s_1 + s_2 + s_3 + \cdots + s_F$ is the total number of all the sides of
all F faces. In this total each edge of the polyhedron is counted exactly
twice (as it is adjacent to exactly two faces) and so

$$s_1 + s_2 + \cdots + s_F = 2E$$

where E stands for the number of the edges of polyhedron. Hence we
obtain

$$(!) \qquad \Sigma\alpha = 2\pi(E - F)$$

We have found here a second expression for $\Sigma\alpha$, but there is an essential difference: the first expression, (?) in subsection (4), was just conjectured, but (!) is proved. If we eliminate $\Sigma\alpha$ from (?) and (!) we obtain the relation

$$(??) \qquad F + V = E + 2$$

which, however, we have not proved and, therefore, we let (??) precede it. In fact, (??) is just as doubtful as (?); through the well-proven (!), each of the two relations (?) and (??) follows from the other, and so they stand and fall together, they are equivalent.

(7) *Verifications*. Both the familiar relation (??) and the less familiar relation (?) were discovered by Euler, who did not know that Descartes found the same relations before him. We know about Descartes' work on this subject from a few brief sentences found among his unpublished manuscripts and printed about a century after Euler's death.[2]

Euler devoted two memoirs and a short remark of a third memoir to the subject.[3] The remark is concerned with the sum of the solid angles in a tetrahedron [this sum depends on the shape, as we have said in subsection (2)]. On the whole, the preceding exposition follows Euler's first memoir where he tells how he was led to his discovery, but offers no formal proof, only a variety of verifications. We wish to follow Euler also in this respect. By collecting the material from our foregoing tables and adjoining to it E, the number of edges, we obtain Table III.

TABLE III

Polyhedron	F	V	E
Tetrahedron	4	4	6
Cube	6	8	12
Octahedron	8	6	12
Dodecahedron	12	20	30
Icosahedron	20	12	30
Tower	9	9	16
n-Prism	$n + 2$	$2n$	$3n$
n-Pyramid	$n + 1$	$n + 1$	$2n$
n-Double-pyramid	$2n$	$n + 2$	$3n$

The conjectured relation (??) is verified throughout Table III, which is gratifying but, of course, does not amount to a proof.

[2] Descartes, *Œuvres*, vol. X, pp. 265–269.
[3] Euler, *Opera Omnia*, ser. 1, vol. 26, pp. XIV–XVI, 71–108, and 217–218.

(8) *Reflections on the result obtained.* In his second memoir, Euler attempted a proof for (??). His attempt failed, however; there is an essential gap in his proof. Yet, in fact, the foregoing considerations have brought us quite close to a proof; we just need to realize how far we have advanced.

Let us try to realize the full meaning of our result (!). Especially, let us see what happens when the polyhedron *varies.* Let us imagine that the polyhedron changes *continuously;* its faces are gradually inclining so that their lines and points of intersection, the edges and vertices of the polyhedron, are continuously changing, yet so that the "general plan," or "morphological structure," of the polyhedron, the connection between its faces, edges and vertices, remains unchanged. Then also the *numbers F, E,* and *V* (of faces, edges, and vertices respectively) remain *unchanged.* Such a change may affect each face angle α individually, but, by virtue of the well-proven (!), it cannot affect the face angles collectively, that is, it must leave $\Sigma\alpha$, the sum of all face angles, unchanged. And here we may see an opportunity to profit by such a change: the change could reduce the proposed polyhedron to some more accessible form for which we might more easily compute the (unchanged!) $\Sigma\alpha$.

Indeed, let us choose one of the faces of the polyhedron as "base." We place this base horizontally and we stretch it (whereas the other faces shrink) so that eventually the whole polyhedron can be orthogonally *projected onto its base;* Fig. 15.8 shows the result (*a*) for a cube and (*b*) for a "general" polyhedron. The result is a collapsed polyhedron, flattened into two superposed polygonal sheets (with the same rim): the lower sheet (the "stretched base") is undivided, the upper sheet is divided into $F - 1$ subpolygons, F being the number of faces of the original polyhedron. We let r denote the number of sides of the polygonal rim enclosing both sheets.

We compute $\Sigma\alpha$ for the flattened polyhedron (we know that it has the same value for the original, unflattened polyhedron). The total sum consists of three parts.

Fig. 15.8. Flattened polyhedra. (*a*) (*b*)

The sum of the angles in the lower sheet (the "stretched base") is $(r - 2)\pi$.

The sum of the angles on the rim of the upper sheet is the same.

There remain the angles in the interior of the upper sheet; they surround the $V - r$ interior vertices and so their sum is $(V - r)2\pi$.

From the three component parts, we obtain

$$\Sigma\alpha = 2(r - 2)\pi + (V - r)2\pi$$
$$= 2\pi V - 4\pi$$

This proves our conjecture (?) and, therefore, also our conjecture (??).

15.7. Scientific method: guess and test

The foregoing examples suggest general insights. Of course, such insights will emerge more spontaneously and with better documentation from more work on these examples and on similar ones (see the Examples and Comments at the end of this chapter). Yet, even what we have discussed may lead to some remarks tending toward the following.

Observation may lead to discovery.

Observation should disclose some regularity, pattern, or law.

Observation has more chance to yield worthwhile results if guided by some good remark, by some insight [as in sect. 15.6(4)].

Observation yields only tentative generalizations, conjectures, not proofs.

Test your conjecture: examine particular cases and consequences.

Any particular case or consequence that is verified (turns out to be true) adds to the credit of the conjecture.

Distinguish carefully between suggestion and proof, between conjecture and fact.

Do not neglect analogies: they may lead to discovery [as the analogy between polygons and polyhedra, sect. 15.6(1)].

Examine extreme cases [such as degenerate triangles and collapsed polyhedra, sect. 15.4 and sect. 15.6(8)].

These remarks would deserve to be stated more precisely, with more details, with more system and, especially, with much broader documentation (*cf.* MPR). Yet even as they stand, as they may emerge from examples such as the preceding and from well-directed classroom discussions, they can give students on the high school level a basic insight into the nature of science. Philosophers, past and present, have offered and do offer widely divergent views about the nature of science, of "scientific method," of "induction," etc. Yet what do scientists actually do? They devise hypothetical explanations and submit their hypotheses to the test of experience. If you want a description of scientific method in three syllables, I propose:

GUESS AND TEST

15.8. Some desirable features of "research problems"

The problems just presented differ from the usual routine problems in several respects. I wish to emphasize just three points.

(1) The schoolboy receives his problem ready made from the textbook or from the teacher and, in many cases, the teacher cares little and the textbook less whether the boy takes any interest or not in the problem. On the contrary, the mathematician's most crucial step may be to *choose* his problem; to spot, to invent a problem that is attractive and worthwhile but not beyond his force. In sect. 15.2 and 15.4 the teacher proceeds so that the students should *have some share in proposing the problem* [*Cf.* sect. 14.5(1)].

(2) Most textbook problems have little connection; they serve to illustrate just one rule and offer some practice in its application. After having rendered this service they may sink into oblivion. Yet the problems of sect. 15.2 and 15.6 have a *rich background;* they suggest challenging problems which suggest further challenging problems, until the ramifications of the original problem cover a wide domain. (Such ramifications will be discussed to some extent in the Examples and Comments at the end of the chapter.)

(3) In many classrooms "guessing" is taboo, whereas in mathematical research "First guess then prove" is almost the rule. In the foregoing problems, observation, conjectures, inductive arguments, in short, *plausible reasoning* play a prominent role.

(4) Although point (1) (participation of the students in proposing the problem) is not unimportant, the other two points are more momentous. Problems with a *background* connected with the world around us, or with other domains of thought, and problems involving *plausible reasoning,* challenging the judgment of the students have more chance to lead them to intellectual maturity than the problems that fill the textbooks and serve only to practice this or that isolated rule.

15.9. Conclusion

As I see them, examples and remarks such as those presented in this chapter are communicable on the high school level and they can do, let us say, three things for the students.

First, they may give the student a taste of mathematics in the making, of independent creative work.

Second, and this is even more important, as it may serve a larger fraction of the student population: they contribute not only to the understanding of mathematics, but also to that of the other sciences. In fact, they give a reasonably good first idea of "inductive research" and "scientific method."

Third, they reveal an aspect of mathematics which is as important as it is rarely mentioned; mathematics appears here as a close relative to the natural sciences, as a sort of "observational science" in which observation and analogy may lead to discoveries. This aspect should especially appeal to prospective users of mathematics, to future scientists and engineers.

And so I hope that mathematical discovery, scientific method, and the inductive aspect of mathematics will not be so completely neglected in the high schools of the future as they are in the high schools of today.

Examples and Comments on Chapter 15

First Part

15.1. Among the several conjectures suggested by the list of sect. 15.2 and stated in sect. 15.3, are there some that you can prove? Pick out some readily accessible assertion and prove it.

15.2. [Sect. 15.4] Devise further means to check Heron's formula.

15.3. [Sect. 15.5] In Fig. 15.5, an arc of the unit circle divides the triangle (the points of which represent triangular shapes) into two parts, one above, the other below, the arc. What is the difference between the shapes represented by those two parts?

15.4. [Sect. 15.6(8)] Try to devise a *more definite* transition from the "general" convex polyhedron to the "flattened" one.

15.5. We consider a convex polyhedron with F faces, V vertices, and E edges. Let

F_n stand for the number of those faces that have precisely n sides, and
V_n for the number of those vertices in which precisely n edges meet.
Tell the value of ΣF_n and that of ΣV_n; the sign Σ is used here as an abbreviation

for $\sum\limits_{n=3}^{\infty}$. (Of course, only a finite number among the F_n are different from 0, and

the same holds for V_n; hence, in fact, Σ denotes a finite sum. The sign Σ will be used in the same sense in some of the following connected problems.)

15.6. (Continued) Express the number of the face angles in several different ways.

15.7. (Continued) By appropriately chosen diagonals ("face diagonals") which do not cross each other, dissect each face of a polyhedron into triangles. Express in several different ways the number of triangles into which the total surface of the polyhedron is so divided.

15.8. Show that

$$E \geqq \frac{3F}{2}, \qquad E \geqq \frac{3V}{2}$$

Can the case of equality be attained in the first inequality, and under what circumstances? Analogous question for the second inequality.

15.9. Show that in any convex polyhedron the average of the face angles is never less than $\pi/3$ but always less than $2\pi/3$.

15.10. Show that in any convex polyhedron there must be a face with less than 6 sides.

15.11. Being given V, the number of vertices of a convex polyhedron, find the maximum of F and the maximum of E. Under what condition are these maxima attained?

15.12. Being given F, the number of faces of a convex polyhedron, find the maximum of V and the maximum of E. Under what condition are these maxima attained?

15.13. When a straight-line segment connects two vertices of a convex polyhedron, there are three possible cases: the connecting segment may be an edge, or a face diagonal, or a diagonal. The last case arises iff, except the two endpoints, no point of the segment lies on the surface of the polyhedron. Let D denote the number of diagonals of the polyhedron; E, F, V, F_n, and V_n will be used in the same meaning as earlier.

(1) Find D for the five regular polyhedra.

(2) Find D for the n-prism, the n-pyramid, and the n-double-pyramid.

(3) Express D in terms of F when all faces of the polyhedron are polygons with the same number of sides $n = 3, 4, 5, \ldots$.

(4) Express D generally.

Illustrate each general case by examples. Be careful; questions may be misleading.

15.14. [Sect. 15.6(2)] We consider a tetrahedron and let $\Sigma\delta$ stand for the sum of its six dihedral angles and $\Sigma\omega$ stand for the sum of its four solid angles.

Compute these two sums in the following three limiting cases:

(1) The tetrahedron collapses into a triangle, three edges become sides of the triangle, three other edges line segments drawn from an interior point of the triangle to its vertices.

(2) The tetrahedron collapses into a convex quadrilateral, its six edges become the four sides and the two diagonals of the quadrilateral.

(3) One vertex of the tetrahedron goes to infinity, the three edges converging to it become parallel and perpendicular to the opposite face.

(A sphere with radius 1 is described about the vertex of a polyhedral angle as center. That part of the surface of the sphere that falls within the polyhedral angle is a spherical polygon. The area of this spherical polygon measures the "solid angle.")

15.15. (Continued) Examine the answer to ex. 15.14. Compare the two sums examined. Do they vary in the same way? Are their variations related?

15.16. For a polyhedron with F faces, V vertices, and E edges, we let $\Sigma\delta$ denote the sum of its E dihedral angles and $\Sigma\omega$ the sum of its V solid angles. Compute the two sums for a cube.

15.17. (Continued) Compute the two sums for two different easily accessible (degenerate) cases of the n-pyramid.

15.18. (Continued) Compute the two sums in accessible (limiting) cases of the n-prism and the n-double-pyramid.

15.19. (Continued) In all cases considered, compare the two sums with F, V, and E; observe the variation of the quantities compared; which variations appear most closely related?

15.20. (Continued) If you have found a rule that is supported by all your observations, try to prove it.

Second Part

15.21. Try to guess the answers to the following questions.
Of all triangles inscribed in a given circle which one has the largest area?
Of all quadrilaterals inscribed in a given circle which one has the largest area?
Of all polygons with a given number n of sides inscribed in a given circle which one has the largest area?

15.22. Try to guess the answers to the following questions.
Of all the triangles circumscribed about a given circle which one has the smallest area?
Of all the quadrilaterals circumscribed about a given circle which one has the smallest area?
Of all the polygons with a given number n of sides circumscribed about a given circle which one has the smallest area?

15.23. *The Principle of Nonsufficient Reason.* The "usual" answers given to ex. 15.21 and 15.22 are correct.[4] We shall not discuss the proof here. We wish to explore why do people so often guess correctly in similar situations.

Of course, we cannot expect to find a very definite answer. Yet I think that the following expresses feelings voiced by many people.

Why are the regular polygons so popular? The circle is the most "perfect," the most symmetric plane figure; it has infinitely many axes of symmetry, it is symmetric with respect to each of its diameters. Of all polygons with a given number of sides the regular polygon is the "nearest in perfection" to the circle: it is the most symmetric, it has more axes of symmetry than the others. And so we expect that the inscribed regular polygon will "fill" the circle better (and the circumscribed regular polygon will "hug" the circle tighter) than any other polygon with the same number of sides.

Analogy plays a role too. The regular polygon yields the extremum in the isoperimetric problem (sect. 15.3, ex. 15.1) which is similar to the foregoing problems.

---

[4] Concerning ex. 15.21 see MPR vol. 1, pp. 127–128.

There are still other plausible arguments. We are going to discuss one that is rather subtle but deserves special attention. We are facing here problems with several unknowns where the condition is the same for each unknown; the condition does not favor any vertex of the polygon more than any other vertex, no side more than any other side. Hence we may expect that, in the polygon that satisfies the condition and so solves the problem, all sides will be equal and all angles will be equal. Thus we expect that the regular polygon will be the solution.

Underlying this expectation is a principle of plausible inference which we may try to express as follows:

"No one should be favored of eligible possibilities among which there is no sufficient reason to choose."

We may term this the "principle of nonsufficient reason." This principle is of some importance in problem solving; quite often it enables us to forecast the solution, or choose the procedure leading to the solution. In a mathematical context, a more specific formulation of the principle may be advantageous:

"Unknowns that play the same role in the condition may be expected to play the same role in the solution."

Or shorter: "No difference in the conditions, no difference in the results." Or: "Expect the same values for unknowns subject to the same conditions."

In geometrical problems, the principle favors symmetry (as we have seen.) Hence we may sometimes find the following formulations of the principle of nonsufficient reason more suggestive (although, in fact, they are less clear):

"We expect that any symmetry found in the data and condition of the problem will be mirrored by the solution."

"Symmetry should result from symmetry."

To some extent, the "symmetry found in the data and condition" should be mirrored not only by the "solving object" but also by the "solving procedure."[5]

Of course, we should not forget that the principle is merely heuristic and should not take plausibility for certainty.[6]

The principle of nonsufficient reason plays a certain role also in not purely mathematical questions.[7]

A striking example which flies in the face of the principle can be concisely described if we use some algebraic terminology. The problem is: Find n quantities of which the n elementary symmetric functions are given. The principle of nonsufficient reason makes us expect that those n quantities will be equal. Yet the n roots of an algebraic equation of which the coefficients are given "at random" should be expected to be all different from each other.

15.24. *The ass of Buridan.* An ass was very hungry when he found himself facing two equal and equally appetizing haystacks, one to his left, the other to his right, and himself in such a position in the middle that there was perfect symmetry.

---

[5] See HSI, pp. 199–200 (Symmetry), and ex. 5.13 for the terminology.

[6] *Cf.* MPR, vol. 1, pp. 186–188, ex. 40 and 41. See also a paper of the author "On the role of the circle in certain variational problems" *Annales Univ. Scient. Budapest. Sectio Math.* v. 3–4, 1960–61, pp. 233–239.

[7] See J. M. Keynes, *A treatise on probability,* pp. 41–64.

Attracted equally by both haystacks the ass could not decide between them and died of hunger.

Poor ass—he fell a victim to the principle of nonsufficient reason.

15.25. Of all polyhedra inscribed in a given sphere and having a given number V of vertices, which one has the largest volume?

Try to guess the answer for $V = 4$, 6, and 8.

15.26. Of all polyhedra circumscribed about a given sphere and having a given number F of faces, which one has the smallest volume?

Try to guess the answer for $F = 4$, 6, and 8.

15.27. Given the radius r of a sphere, compute the volume of the inscribed cube.

15.28. Consider the sphere with radius r as the globe. Inscribe a regular hexagon in the equator. The six vertices of this hexagon, the north pole, and the south pole are the eight vertices of a double pyramid. Compute its volume.

Any remarks?

15.29. Given the radius r of a sphere, compute the volume of the circumscribed regular octahedron.

15.30. A right prism with regular hexagonal base is circumscribed about a sphere with radius r, which we regard as the globe. The surface of the prism touches the sphere in six points equally spaced along the equator, in the north pole, and in the south pole. Compute the volume of the prism.

Any remarks?

15.31. Compare the solids considered in the foregoing (ex. 15.27 with ex. 15.28, ex. 15.29 with ex. 15.30) and try to find a plausible explanation.

15.32. Here is a plausible assumption: of two polyhedra inscribed in the same sphere and having the same number V of vertices, the one with more faces and edges has more chance to "fill" the sphere. Granted this, what kind of polyhedron do you expect to be the solution of ex. 15.25?

15.33. Here is a plausible assumption: of two polyhedra circumscribed about the same sphere and having the same number F of faces, the one with more vertices and edges has more chance to "hug" the sphere tighter. Granted this, what kind of polyhedron do you expect to be the solution of ex. 15.26?

15.34. Are there any more remarks suggested by ex. 15.31?

15.35. Of all polyhedra having a given surface area and a given number F of faces, which one has the largest volume?

Try to guess the answer for $F = 4$, 6, and 8.

15.36. Solve the system, giving all solutions:

$$2x^2 - 4xy + 3y^2 = 36$$
$$3x^2 - 4xy + 2y^2 = 36$$

What about the principle of nonsufficient reason?

Fig. 15.9. Equal weights at equal distances.

15.37. Solve the system, giving all solutions:

$$6x^2 + 3y^2 + 3z^2 + 8(yz + zx + xy) = 36$$
$$3x^2 + 6y^2 + 3z^2 + 8(yz + zx + xy) = 36$$
$$3x^2 + 3y^2 + 6z^2 + 8(yz + zx + xy) = 36$$

15.38. Solve the system, giving all solutions:

$$x^2 + 5y^2 + 6z^2 + 8(yz + zx + xy) = 36$$
$$6x^2 + y^2 + 5z^2 + 8(yz + zx + xy) = 36$$
$$5x^2 + 6y^2 + z^2 + 8(yz + zx + xy) = 36$$

15.39. *The principle of nonsufficient reason in physics. Or, nature ought to be predictable.* The beginning of Archimedes' work "On the Equilibrium of Planes or the Centres of Gravity of Planes"[8] deals with the equilibrium of the lever. (The lever is a rigid horizontal bar supported at one point which is called the fulcrum; the weight of the bar is regarded as negligible.) Archimedes considers the case where the midpoint of the lever is the fulcrum and the two endpoints carry equal weights, see Fig. 15.9; he admits as evident that in this situation of perfect symmetry there is equilibrium; in fact, his first postulate states: "Equal weights at equal distances are in equilibrium." Indeed, the lever is in the situation of Buridan's ass: it has no sufficient reason to incline to one side rather than to the other.

We can penetrate here a little deeper. Let us see what happens if somebody contradicts Archimedes' postulate and proposes a different rule: in the situation of Fig. 15.9 the right-hand weight will sink. Well, if this prediction turns out correct for me as I am looking at the lever, it must turn out wrong for my friend who, facing me, is looking at the lever from the opposite side: the rule which contradicts Archimedes' postulate *cannot be generally valid.* We may now perceive a deeper source of the confidence with which we are inclined to accept Archimedes' postulate: we want nature to be predictable.

15.40. *Choosing n points on a spherical surface.* We restate ex. 15.25 and ex. 15.26 as the first two in a series of analogous problems.

On the surface of a given sphere choose n points so that

(1) the *inscribed polyhedron* of which the vertices are the n points should have *maximum volume;*

(2) the *circumscribed polyhedron* the n faces of which touch the sphere in the n points should have *minimum volume;*

(3) the shortest of the $n(n - 1)/2$ distances between the n points should

---

[8] See his Works, edited by T. L. Heath, p. 189.

be a maximum (a "maximum minimorum": this is the "problem of the *n* misanthropes");

(4) each point carrying a unit electric charge, the *n* mutually repelling charges should be in the *most stable electrostatic equilibrium*.

(5) On the surface of the sphere, there is an arbitrary continuous mass distribution the density of which is observed at *n* points. Choose the *n* points so that on the basis of those *n* observations the total mass can be most accurately estimated. (This is the "problem of the *n reporters*" of a worldwide news agency, or the problem of *best interpolation*. For a line segment, Gauss has solved the analogous problem, in a certain sense, by his celebrated mechanical quadrature.)

In all five foregoing problems, if *n* = 4, 6, 8, 12, or 20, the vertices of a certain inscribed regular polyhedron deserve consideration, although they may not yield the solution as certain foregoing examples show. *Cf.* L. Fejes-Tóth, *Lagerungen in der Ebene, auf der Kugel und im Raum.*

If the *n* points are chosen at random (the *n* brightest fixed stars on the sky appear to be so chosen when *n* is not very large), the average distance of a point from its next neighbor, its second next neighbor, its third next neighbor, and so on, can be computed. See a paper of the author, *Vierteljahrsschrift der Naturforschenden Gesellschaft in Zürich,* v. 80, 1935, pp. 126–130.

Third Part

15.41. *More problems.* Consider some research problems similar to, but different from, the problems discussed in this chapter. Pay especial attention to such questions as the following: Does the problem fit, and where does it fit, into the curriculum? Is the problem instructive? Has it a worthwhile background? Does it illustrate some idea of general interest? Does it afford opportunity for inductive, plausible reasoning? Or opportunity for a challenging proof on the level of the class? How should it be presented to the class?

15.42. In sect. 15.4 we tested a general formula by discussing particular cases. Where have you opportunity for a discussion of this kind? Carry through such a discussion in a few more cases. What is the merit of such a discussion?

15.43. The main purpose of sect. 15.5 is to illustrate graphically an aspect of inductive reasoning. Could the students benefit from this section in other respects?

15.44. *Periodic decimals.* The three decimal expansions

$$\tfrac{1}{6} = 0.166666666\ldots$$
$$\tfrac{1}{7} = 0.142857142\ldots$$
$$\tfrac{1}{8} = 0.125$$

belong to three different types. The one representing $\tfrac{1}{8}$ is a *terminating* decimal expansion, the two others are *infinite*. In fact, they are recurring, repeating, or *periodic decimals;* with the standard notation, they are written in the form

$$\tfrac{1}{6} = 0.1\overline{6}$$
$$\tfrac{1}{7} = 0.\overline{142857}$$

The bar is placed over the repetend or *period,* that is, the sequence of digits that must be infinitely often repeated. The period of $\frac{1}{6}$ is of length 1, that of $\frac{1}{7}$ of length 6; generally, the number of digits in a period is called its *length.* The decimal expansion of $\frac{1}{7}$ is *purely periodic* that for $\frac{1}{6}$ is *mixed:* The first does not and the second does contain an initial sequence of digits that does not belong to the period. Here is one more example for each of the three types

$$\tfrac{39}{44} = 0.88\overline{63} \qquad \tfrac{19}{27} = 0.\overline{703} \qquad \tfrac{19}{20} = 0.95$$

Find out by observation as much as you can about the three types of decimal expansions, about the length of their periods, about the distribution of the digits in the period, or whatever else you find worth noticing. Try to prove or disprove the guesses to which observation leads you.

Choose the fractions you wish to observe, or look at the decimal fractions into which the numbers mentioned under (1) through (7) are expanded.

(1) $\frac{1}{7}, \frac{2}{7}, \frac{3}{7}, \frac{4}{7}, \frac{5}{7}, \frac{6}{7}, \frac{7}{7}$

(2) $\frac{1}{7}, \frac{10}{7}, \frac{100}{7}, \ldots$

(3) All proper fractions with a denominator that is relative prime to the numerator and is less than 14.

(4) All proper fractions with denominator 27 whose numerator is relative prime to 27.

(5) $\frac{1}{3}, \frac{1}{7}, \frac{1}{11}, \frac{1}{37}, \frac{1}{41}, \frac{1}{73}, \frac{1}{101}, \frac{1}{239}$

(6) $\frac{1}{9}, \frac{1}{99}, \frac{1}{999}, \frac{1}{9999}$

(7) $\frac{1}{11}, \frac{1}{101}, \frac{1}{1001}, \frac{1}{10001}$

At any rate observe and try to understand thoroughly that

$$7.00000\ldots = 6.99999\ldots$$

$$0.50000\ldots = 0.49999\ldots$$

15.45. (Continued) Observe that

$\frac{1}{9} = 0.111111\ldots$	$\frac{1}{11} = 0.090909\ldots$
$\frac{1}{27} = 0.037037\ldots$	$\frac{1}{37} = 0.027027\ldots$
$\frac{1}{99} = 0.01010101\ldots$	$\frac{1}{101} = 0.00990099\ldots$
$\frac{1}{271} = 0.0036900369\ldots$	$\frac{1}{369} = 0.0027100271\ldots$

and EXPLAIN.

15.46. (Continued) Starting from decimal expansions and passing from the base 10 to the base 2 we arrive at binary expansions. Here is an example:

$$\tfrac{1}{3} = 0.01010101\ldots$$

if we interpret the right-hand side in the binary system, that is, *if* we take the foregoing equation in the meaning

$$\frac{1}{3} = \frac{1}{2^2} + \frac{1}{2^4} + \frac{1}{2^6} + \frac{1}{2^8} + \cdots$$

Examine binary expansions in the same way as decimal expansions have been examined in ex. 15.44 and 15.45.

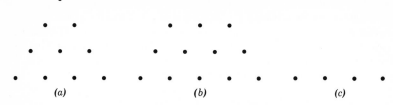

Fig. 15.10. Triangular and trapezoidal numbers.

15.47. (Continued) Assess the educational merit of the project of research sketched in ex. 15.44, 15.45, and 15.46.

15.48. *Trapezoidal numbers:* Fig. 15.10a represents the triangular number

$$1 + 2 + 3 + 4 = 10$$

cf. ex. 3.38 and Fig. 3.8. We could call analogously the number

$$3 + 4 + 5 = 12$$

represented by Fig. 15.10b a "trapezoidal" number. If we include extreme cases in our definition (which is often desirable) we must regard also the numbers represented by (a) and (c) in Fig. 15.10 as "trapezoidal." Yet then each positive integer would be "trapezoidal" (since it can be represented by one row of points as in Fig. 15.10c) and the definition would be pointless. Thus we are led to the following definition.

Let $t(n)$ denote the number of *trapezoidal representations* of the positive integer n, that is, the number of representations of n as sum of consecutive positive integers. Here are a few examples:

$$6 = 1 + 2 + 3$$
$$15 = 7 + 8 = 4 + 5 + 6 = 1 + 2 + 3 + 4 + 5$$

When

$$n = 1, 2, 3, 6, 15, 81, 105$$
$$t(n) = 1, 1, 2, 2, \ \ 4, \ \ 5, \ \ \ 8$$

Find a "simple expression" for $t(n)$ by appropriate observation followed, if possible, by a proof.

15.49. (Continued) Fig. 15.11 offers graphic help to survey your observations. Let us call the expression of n in the form

$$n = a + (a + 1) + (a + 2) + \cdots + (a + r - 1)$$

(a sum of r terms) a trapezoidal representation of n *with r rows.* Iff n admits a trapezoidal representation with r rows, the point with abscissa n and ordinate r is marked with a dot in Fig. 15.11.

If $t(n) = 1$, the only trapezoidal representation of n must be the "trivial" representation with $r = 1$. List the numbers n in Fig. 15.11 for which $t(n) = 1$.

What is $t(p)$ if p is a prime number?

15.50. (Continued) Let $s(n)$ denote the number of representations of the positive integer n as sum of consecutive *odd* positive integers.   Find an expression for $s(n)$.

Examples

$$15 = 3 + 5 + 7$$
$$45 = 13 + 15 + 17 = 5 + 7 + 9 + 11 + 13$$
$$48 = 23 + 25 = 9 + 11 + 13 + 15 = 3 + 5 + 7 + 9 + 11 + 13$$

When

$$n = 2, 3, 4, 15, 45, 48, 105$$
$$s(n) = 0, 1, 1, \quad 2, \quad 3, \quad 3, \quad 4$$

15.51. Assess the project of research sketched in ex. 15.48 and 15.49.

15.52. Consider three plane figures:
(1) a square with a vertical diagonal,
(2) the circle (with radius a) circumscribed about it,
(3) the square circumscribed about the circle with a vertical side.
The vertical diagonal of (1) divides each figure in two symmetric *halves*. Rotating about their common vertical axis of symmetry, the three plane figures describe three solid figures:
(1) a double cone,
(2) a sphere,
(3) a cylinder.
Compute, in all three cases,
    V, the volume of the solid figure,
    S, the surface area of the solid figure,
    A, the area of the plane figure,
    L, the perimeter of the plane figure,
    X_A, the distance of the center of gravity of the half-area of the plane figure from the axis of revolution,
    X_L, the distance of the center of gravity of the half perimeter of the plane figure from the axis of revolution.
Group the eighteen quantities in a 3×6 table, OBSERVE, and try to explain what you have observed.

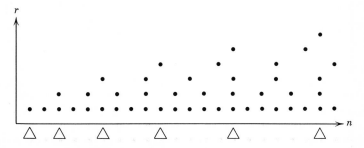

Fig. 15.11. Trapezoidal representation of n with r rows.

15.53. Observe that

$$\sqrt{2} - 1 \qquad\qquad\qquad = \sqrt{2} - \sqrt{1}$$
$$(\sqrt{2} - 1)^2 = 3 - 2\sqrt{2} \quad = \sqrt{9} - \sqrt{8}$$
$$(\sqrt{2} - 1)^3 = 5\sqrt{2} - 7 \quad = \sqrt{50} - \sqrt{49}$$
$$(\sqrt{2} - 1)^4 = 17 - 12\sqrt{2} = \sqrt{289} - \sqrt{288}$$

try to generalize, and prove your guess.

15.54. Quite often it matters little what your guess is, but it always matters a lot how you test your guess.

15.55. *Fact and conjecture.* The following story of which I cannot guarantee the authenticity is about Sir John and a janitor. Sir John, who is a member of the Royal Society, can be expected to draw a sharp distinction between a fact and a conjecture; the janitor, who is employed in the Royal Society building, unexpectedly drew such a distinction on a certain occasion.

One day Sir John arrived somewhat late to a meeting of the Royal Society and was visibly in a hurry. He had to deposit his hat in the cloakroom and receive a check with a number for it. The janitor who was in charge of the cloakroom that day said obligingly: "Please, do not wait for the check, Sir, I shall give you your hat without a check." Sir John went thankfully to the meeting without the check, but was somewhat worried about the fate of his hat. When, however, he entered the cloakroom after the meeting, the janitor, without hesitation, gave him his hat. Sir John was visibly pleased, and I do not know what prompted him to ask the janitor: "How did you know that this is my hat?" I do not know either what happened to the janitor—perhaps he found Sir John's tone unduly patronizing— at any rate he answered rather sharply: "Sir, I do not know whether it is your hat; it is the hat you gave me."

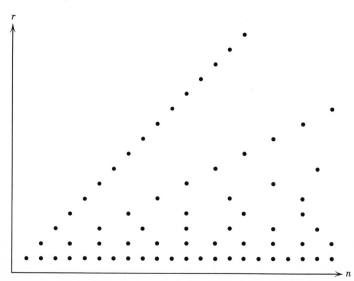

Fig. 15.12. For the observant reader, a diagram due to Leibnitz: *n* is divisible by *r*.

SOLUTIONS

Chapter 7

7.1. The connecting lines exhibited by Fig. 7.10 indicate the relation

$$F = Q + 4T + 4P$$

Obviously

$$Q = a^2h, \qquad T = \frac{h}{2} \cdot \frac{b-a}{2} \cdot a, \qquad P = \frac{h}{3}\left(\frac{b-a}{2}\right)^2$$

Introduce the connecting lines indicating these relations into Fig. 7.10 and satisfy yourself that the foregoing relations yield the same final expression for F as the approach considered in the text.

7.2. See Fig. S7.2, (3) and (4); the latter represents the mental situation just before the dramatic last line in Volume I, p. 36.

No solution: **7.3, 7.4, 7.5.**

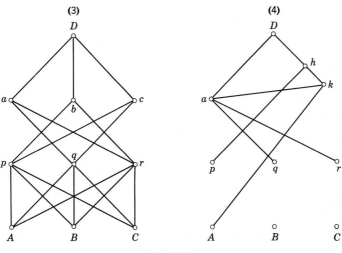

Fig. S7.2.

Chapter 8

8.3. (Stanford 1956.)　From Fig. 8.3

$$AB' = \frac{b}{\sqrt{3}}, \qquad AC' = \frac{c}{\sqrt{3}}$$

Apply the law of cosines to $\triangle B'AC'$:

$$3x^2 = b^2 + c^2 - 2bc \cos\left(\alpha + \frac{\pi}{3}\right)$$

Apply the law of cosines to $\triangle ABC$ to express $bc \cos \alpha$, set $bc \sin \alpha = 2T$ where T is the area of $\triangle ABC$, and obtain

$$6x^2 = a^2 + b^2 + c^2 + 4\sqrt{3}\,T$$

T is symmetric in a, b, and c.

8.4. The cross-section is

(1) a rectangle with area $2yz = \dfrac{2hx\sqrt{a^2 - x^2}}{a}$,

(2) a right triangle with area $\dfrac{xz}{2} = \dfrac{h(a^2 - y^2)}{2a}$,

(3) a segment of a circle.

Plan (2) is preferable (rational function of y) and the desired volume equals

$$\tfrac{1}{2}\int_{-a}^{a} xz\, dy = \frac{2a^2h}{3}$$

No solution: **8.1, 8.2, 8.5, 8.6, 8.7, 8.8.**

Chapter 9

9.5. See sect. 4.4.　The theorem A, stated in sect. 4.4(1) and illustrated by Fig. 4.4 is proved by the help of the weaker theorem B, stated in sect. 4.4(2) and illustrated by Fig. 4.3.

9.6. The problems A and B are equivalent.　There is a distinct advantage in passing from A to B since we arrive so at smaller numbers.　Repeating this passage, we obtain successively the following pairs of numbers:

$$(437, 323) \qquad (323, 114) \qquad (209, 114) \qquad (114, 95) \qquad (95, 19)$$
$$(76, 19) \qquad (57, 19) \qquad (38, 19) \qquad (19, 19)$$

Thus all the common divisors of 437 and 323 are 1 and 19, the divisors of 19 which is their greatest common divisor.　The procedure used in this example is generally applicable, is of basic importance, and is called *Euclid's algorithm* (see Euclid VII 2).

9.7. (1) Under certain important and frequently arising circumstances the condition of B is wider than that of A.　(Here is a simple case: $f(x)$ is defined, continuous, and possesses a derivative in the closed interval $a \leqq x \leqq b$; moreover it is known that the maximum of $f(x)$ is not attained for $x = a$ nor for $x = b$.)

(2) Finding the roots of the equation $f'(x) = 0$ appears in most cases as the more familiar problem; moreover we know means to remove those roots which do not yield the maximum of $f(x)$.

9.8. (1) Theorem B is stronger; it immediately implies A.

(2) B is easier to prove than A, because B adds to A a precise detail on which we can start to work: if we have only the less complete statement A, we have still to discover that precise detail, or some equivalent fact; the stronger theorem B is more accessible than A, because it is more explicit than A. This is a typical case (*cf.* HSI, pp. 114–121, especially p. 121, Induction and mathematical induction, especially point 7).

(3) Let A denote any vertex of the triangle and M the midpoint of the opposite side; prove first that $\triangle MGO \sim \triangle AGE$, from which $MO \parallel AE$ follows.

9.9. (1) A is a problem to prove, B is a problem to find which is more ambitious than A: we can foresee from the outset that a complete solution of B will either prove or disprove the assertion stated by A.

(2) A is a problem on limits, B on algebraic inequalities, and so B appears more elementary.

(3) We omit the case $\epsilon \geqq 1$ which is easier. In the case $\epsilon < 1$ we have a chain of equivalent conditions:

$$\sqrt{x+1} < \epsilon + \sqrt{x}$$
$$1 < \epsilon^2 + 2\epsilon\sqrt{x}$$
$$\left(\frac{1-\epsilon^2}{2\epsilon}\right)^2 < x$$

That is, from a certain value of x onward, the quantity $\sqrt{x+1} - \sqrt{x}$ is less than ϵ, an arbitrary (arbitrarily small!) positive quantity. This proves A.

9.10. (1) The two propositions with which A and B are concerned are equivalent (by contraposition) and so are the two problems A and B themselves.

(2) The statement, "n is composite," affirms the existence of two integers a and b such that $n = ab, a > 1, b > 1$. The statement "n is prime" denies that n is composite (we can disregard here the case $n = 1$) and this "negative" statement offers "less foothold." Hence B appears more tractable.

(3) If $n = ab$

$$2^n - 1 = (2^a)^b - 1$$

is divisible by $2^a - 1$.

9.11. Set, for $m = 1, 2, 3, \ldots,$

$$a_{3m-2} = 2m^{-1/3}, \qquad a_{3m-1} = a_{3m} = -m^{-1/3}$$

For a generalization of the present problem see *American Mathematical Monthly,* vol. 53, 1946, pp. 283–284, problem 4142. It seems impossible to set up helpful additional requirements, like (III) and (IV), without some idea, without some premonition or anticipation of the solution.

9.12. If $p_1, p_2, \ldots, p_l$ are different primes, and $n = p_1{}^{a_1}p_2{}^{a_2} \cdots p_l{}^{a_l}$ then

$$\tau(n) = (a_1 + 1)(a_2 + 1) \cdots (a_l + 1)$$

9.13. Ex. 1.47 and sect. 1.3(1); sect. 2.5(2), 2.5(3); sect. 3.2 and sect. 3.1.

9.15. *Why?* Here are two more tasks analogous to ex. 9.6–9.10.
(I) Compare the two problems:

A. Find $\sqrt[4]{100}$.

B. Find $\sqrt{100}$.

†(II) Let $f(x)$ and $g(x)$ stand for some given functions. Compare the two problems:

A. Prove that $\int_a^b f(x)\, dx \geqq \int_a^b g(x)\, dx$.

B. Prove that $f(x) \geqq g(x)$ for $a \leqq x \leqq b$.

Wherefrom? Auxiliary problems that originate, overtly or covertly, from one or the other of the four sources named seem to be the most usual. An instructive example: generalization, specialization, and analogy contribute jointly to the solution of ex. 3.84; see Vol. 1, p. 97, p. 191, and HSI ex. 20, pp. 238, 241–242, 252–253.

No solution: **9.1, 9.2, 9.3, 9.4, 9.14.**

Chapter 10

10.2.

(1)

```
_ _ _ E _ _ _ _ _ _
_ _ _ E _ _ _ A _ _
_ W _ E _ _ _ A _ _
_ W _ E _ H _ A _ _
_ W _ E _ H _ A _ T
```

(2)

$$\int \sqrt{x^7 - x^4}\, dx =$$
$$\int x^2 \sqrt{x^3 - 1}\, dx =$$
$$\int \sqrt{x^3 - 1} \cdot x^2 dx =$$
$$\tfrac{1}{3}\int \sqrt{x^3 - 1} \cdot 3x^2\, dx =$$
$$\tfrac{1}{3}\int (x^3 - 1)^{1/2}\, d(x^3 - 1)$$

No solution: **10.1.**

Chapter 11

11.2. Fig. 1.1 to 1.6; Fig. 7.1 to 7.6, summarized by Fig. 7.8.

11.3. See the two examples quoted in the solution of ex. 11.2: in the first, it may be Fig. 1.6; in the second, it may be Fig. 7.3.

11.4. Ex. 10.2.

11.5. Recognizing in Fig. 4.2 the abscissas $x_1, x_3, \ldots, x_n$ as *roots* of the desired polynomial $f(x)$.

11.6. The transition from Fig. 1.11 to Fig. 1.12. Or, in MPR, vol. II, p. 144, the transition from Fig. 16.2 to Fig. 16.3.

No solution: **11.1, 11.7, 11.8, 11.9, 11.10, 11.11.**

Chapter 12

12.4.
(1) The point P lies on two given straight lines; by construction, . . .
The point P lies on a given straight line, and on a given circle; by construction, ...
The point P lies on two given circles; by construction, ...
(2) Given the lengths of two sides of $\triangle ABC$, and the included angle, the third side being AB, ...
Given the length of the side BC and two angles of $\triangle ABC$, ...
Given, in the right triangle $\triangle ABC$, the lengths of the two legs AC and BC, ...
(3) Given the base and the height, ...
Given three sides of $\triangle ABC$, ...
[Also all sets of data listed under (2)].
(4) Given the area of the base $\triangle ABC$ and the length of the altitude from the vertex D, ...

12.5.
(1) If $\triangle ABC \cong \triangle EFG$, ...
If $ABFE$ is a parallelogram, ...
(2) If $\triangle ABC \sim \triangle EFG$, ...
If $\angle ABC$ and $\angle EFG$ are corresponding angles, obtained by cutting two parallels by a transversal, ...
If $\angle ABC$ and $\angle EFG$ are inscribed in the same circle and intercept the same arc, ...
And so on; see ex. 8.8.
(3) If the figure $ABCD$ . . . is similar to the figure $EFGH$ . . . , the points being mentioned in the order as they correspond to each other, . . .
If $AE \parallel BF \parallel CG \parallel DH$, and A, B, C, D are collinear and also E, F, G, H are collinear . . . (A less heavy statement of the theorem is the following: "Corresponding parts of straight lines cut by parallels are proportional.")
(4) If, in $\triangle ABC$, $\angle C < \angle B$, . . .

12.6. See ex. 1.47, sect. 2.5(2), ex. 2.10, ex. 2.13, ex. 2.51, sect. 14.6(3), also HSI pp. 37–46, especially p. 38 (Analogy, especially sect. 3) and MPR, vol. 1, pp. 45–46. There are many others.

12.7. A median of a triangle that ends in the midpoint of the side AB, passes through the midpoint of any segment intercepted by the two other sides on a line parallel to AB.
(In the tetrahedron, the cross sections considered are parallelograms. The plane passing through an edge of the tetrahedron and the midpoint of the opposite edge yields a useful intersection.)

12.8. A median of a triangle bisects its area.

(For the tetrahedron, the theorem stated follows from the principle of Cavalieri, since the plane in question bisects the area of each cross section considered in ex. 12.7.)

12.9. The bisector of an interior angle of a triangle divides the opposite side into segments that are proportional to the adjacent sides.

No solution: **12.1, 12.2, 12.3, 12.10, 12.11.**

Chapter 13

No solution: **13.1, 13.2, 13.3.**

Chapter 14

14.1. Approximately $9\frac{2}{3}$ minutes after noon. The longitude of the "central" meridian for Western Standard Time is 120° west.

14.2. Since

$$BD = BD_1, \qquad CD = CD_1$$

the point D belongs to the intersection of two spheres, one with center B and radius BD_1, the other with center C and radius CD_1. The plane of this circle of intersection is perpendicular to the line BC joining the centers of the spheres, and therefore it is perpendicular to the horizontal plane of $\triangle ABC$. Hence the orthogonal projection F of D onto the horizontal plane lies on the line through D_1 perpendicular to BC. Of course, D_2 and D_3 are analogously situated.

14.3. The point F is the *radical center* (defined in textbooks of analytic geometry) of the three circles indicated by short arcs in Fig. 14.1, and D_1F, D_2F, and D_3F are the three radical axes meeting in F.

14.4. Action and perception in sect. 14.6(4).

14.11. The following answers are sketchy.

(7a) Using the law of cosines both of ordinary and of spherical trigonometry, we are led to prove the inequality

$$\frac{b^2 + c^2 - a^2}{2bc} > \frac{-\cos b \cos c + \cos a}{\sin b \sin c}$$

in supposing that $0 < a < \pi, 0 < b < \pi, 0 < c < \pi$, and that the three segments of length a, b, and c, respectively, form a triangle. The inequality can be derived by appropriate manipulation from the fact that the function $(\sin x)/x$ steadily decreases as x increases from 0 to π, and this fact itself can be made plausible by geometric considerations.

(7b) By continuity, a "very small" spherical triangle is "almost plane," and so it is "almost congruent" to its image in the plane, having the same sides; hence the corresponding angles are "almost" equal.

(7c) The stereographic projection of the sphere (from the north pole onto the plane of the equator) preserves the angles.

(7d) The theorem of Archimedes on the surface area of a spherical zone yields a simple area preserving mapping.

(7e) Projection of a hemisphere onto a plane from the center of the sphere.

14.24. The desired remainder is a polynomial of degree not exceeding 1, of the form $ax + b$. Take the problem as solved, suppose that the quotient $q(x)$ of the division has been found. Then you have the identity:

$$x^3 + x^5 + \cdots + x^{17} + x^{19} = (x^2 - 1)q(x) + ax + b$$

which yields, for $x = 1$ and $x = -1$, two equations to determine the two unknowns a and b:

$$7 = a + b, \qquad -7 = -a + b$$

Hence $b = 0$, $a = 7$, and the desired remainder is $7x$.

The prime numbers 3, 5, . . . 17, 19 are conspicuous but irrelevant; if they are replaced by any seven *odd* positive integers the result remains the same. We see this clearly *after* having solved the problem, yet when we are looking at the problem the first time, those prime numbers could lead us on a wild goose chase.

14.25.

(1) πt^2,

(2) $\pi t^2/4$.

No solution: **14.5–14.10, 14.12–14.23, 14.26, 14.27, 14.28.**

Chapter 15

15.1. "Being given the perimeter, find the figure with maximum area." This is the "isoperimetric problem"; it can be proposed for various classes of figures; here are a few references:

triangles: MPR, vol. 1, p. 133, ex. 16;

rectangles: HSI, pp. 100–102, Examine your guess 2;

quadrilaterals: MPR, vol. 1, p. 139, ex. 41;

polygons with a given number of sides: MPR, vol. 1, p. 178;

all plane figures: MPR, vol. 1, pp. 168–183.

For an introduction to some ideas of Jacob Steiner and to connected physical problems see *Modern Mathematics for the Engineer*, second series, edited by E. F. Beckenbach, pp. 420–441.

15.2. Symmetry in a, b, and c; test by dimensions.

15.3. Acute triangles above, obtuse triangles below, the arc.

15.4. Central projection of the polyhedron (of its inside) onto one of its faces w (the "window"); choose as center of projection a point outside the polyhedron, but sufficiently near to an inner point of w. *Cf.* MPR, vol. 1, p. 53, ex. 7.

15.5. $\Sigma F_n = F, \qquad \Sigma V_n = V$

15.6. $\Sigma n F_n = \Sigma n V_n = 2E$

15.7. Use ex. 15.5, ex. 15.6, the definition of sect. 15.6(2), and the final result of sect. 15.6(8)

$$\sum(n - 2)F_n = 2E - 2F = \sum \frac{\alpha}{\pi} = 2V - 4$$

15.8. From ex. 15.6 and 15.5

$$2E = \Sigma n F_n \geqq 3\Sigma F_n = 3F$$
$$2E = \Sigma n V_n \geqq 3\Sigma V_n = 3V.$$

There is equality in the first line iff each face is a triangle, and in the second line iff just three edges end in each vertex.

15.9. *First proof.* Take here for granted (but prove afterwards for yourself) the lemma.

Lemma. If a set of quantities can be subdivided into nonoverlapping subsets so that the average of the quantities in each subset is less than a, then the average of the quantities in the whole set is also less than a.

This lemma remains valid if the relation "less than" expressed by the sign $<$ is replaced by any of the relations expressed by $\leqq, >, \geqq$. We apply the lemma twice, to (1) and (2).

(1) The average of the angles in a face with n sides is

$$\frac{(n-2)\pi}{n} = \left(1 - \frac{2}{n}\right)\pi \geqq \frac{\pi}{3}$$

(2) The sum of the face angles with common vertex is $<2\pi$, their number $\geqq 3$, and so their average is $<2\pi/3$.

Second proof. By sect. 15.6(6) and ex. 15.6 the average of all face angles is

$$\frac{\Sigma\alpha}{2E} = \frac{2\pi(E-F)}{2E} = \pi\left(1 - \frac{F}{E}\right) \geqq \frac{\pi}{3}$$

by ex. 15.8. Equality is attained iff all faces are triangles.

On the other hand, by Euler's theorem proved in sect. 15.6(8),

$$\frac{\Sigma\alpha}{2E} = \frac{2\pi V - 4\pi}{2E} = \frac{\pi V}{E} - \frac{2\pi}{E} \leqq \frac{2\pi}{3} - \frac{2\pi}{E}$$

by ex. 15.8.

15.10. *First proof.* The average of the angles in a face with n sides is

$$\frac{(n-2)\pi}{n} = \left(1 - \frac{2}{n}\right)\pi \geqq \frac{2\pi}{3}$$

if $n \geqq 6$. If all faces had six sides or more, the average of all face angles would be $\geqq 2\pi/3$, which is impossible by ex. 15.9.

Second proof. From Euler's theorem, ex. 15.8 and ex. 15.6,

$$12 = 6F - 2E + 6V - 4E$$
$$\leqq 6F - 2E$$
$$= \Sigma(6-n)F_n$$
$$12 \leqq 3F_3 + 2F_4 + F_5$$

and so at least one of the three numbers F_3, F_4, and F_5 must be positive.

15.11. (1) If there is a face that has n sides where $n > 3$, we can divide it into $n-2$ triangles by diagonals, and so we can replace it by $n-2$, (which is >1) triangular faces without changing V. Therefore, F cannot be a maximum unless all F faces are triangles.

(2) From 15.8, $2E \geq 3F$ with equality iff $F_4 = F_5 = \cdots = 0$, $F = F_3$. Now,

$$F + V = E + 2 \geq \tfrac{3}{2}F + 2$$
$$V \geq \tfrac{1}{2}F + 2$$
$$F \leq 2(V - 2)$$

and

$$E = F + (V - 2) \leq 3(V - 2)$$

with equality iff each face is a triangle.

15.12. By analogy, both solutions of ex. 15.11 apply (with proper interpretation) to the present case and yield

$$V \leq 2(F - 2), \qquad E \leq 3(F - 2)$$

with equality iff just three edges start from each vertex of the polyhedron.

The inequalities derived have interesting applications. For instance, we combine the second inequality just obtained with ex. 15.8 as follows:

$$\frac{E + 6}{3} \leq F \leq \frac{2E}{3}$$

For $E = 6$ this yields

$$4 \leq F \leq 4$$

that is, the case of the tetrahedron. For $E = 7$ it yields, however,

$$\tfrac{13}{3} \leq F \leq \tfrac{14}{3}$$

and so F cannot be an integer! We are so driven to the conclusion that there exists no convex polyhedron with 7 edges—a fact already noticed by Euler.

15.13.

(1)   0   4   3   100   36
for   tetra-, hexa-, octa-, dodeca-, icosahedron
respectively.

(2) $n(n - 3)$ 0 $1 + \dfrac{n(n - 3)}{2}$

for the n-prism, n-pyramid, n-double pyramid
respectively.

(3) $\dfrac{(F - 2)(F - 4)}{8}$, $\dfrac{(F^2 - 5F + 2)}{2}$, $\dfrac{(9F^2 - 42F + 8)}{8}$

for $n =$ 3,   4,   5

respectively; $n > 5$ is impossible, see ex. 15.10; the ... after 5 is misleading.

(4) In terms of F_n, cf. ex. 15.6, 15.7,

$$D = \frac{V(V - 1)}{2} - E - \sum \frac{n(n - 3)}{2} F_n$$
$$= 1 - \tfrac{1}{4}\Sigma(2n - 3)(n - 2)F_n + \tfrac{1}{8}[\Sigma(n - 2)F_n]^2$$

15.14.

	(1)	(2)	(3)
$\Sigma\delta$	3π	2π	$\dfrac{5\pi}{2}$
$\Sigma\omega$	2π	0	π

15.15. In the cases observed in ex. 15.14, both sums change in the same direction—the change of $\Sigma\omega$ is double of the change of $\Sigma\delta$—and

$$2\Sigma\delta - \Sigma\omega = 4\pi$$

Is this relation true for all tetrahedra?

15.16. *Cf.* ex. 15.19.

15.17. Generalize the cases (1) and (3) of ex. 15.14; in both the result is the same. *Cf.* ex. 15.19.

15.18. *Cf.* ex. 15.19.

15.19.	$2\Sigma\delta - \Sigma\omega$	F	V	E
Tetrahedron	4π	4	4	6
Cube	8π	6	8	12
n-Pyramid	$(2n - 2)\pi$	$n + 1$	$n + 1$	$2n$
n-Prism	$2n\pi$	$n + 2$	$2n$	$3n$
n-Double-pyramid	$(4n - 4)\pi$	$2n$	$n + 2$	$3n$

Neither V nor E, only F increases consistently when $2\Sigma\delta - \Sigma\omega$ increases.

15.20. $2\Sigma\delta - \Sigma\omega = 2\pi F - 4\pi$

For a proof, express the spherical area (the solid angle) associated with a vertex in terms of the dihedral angles associated with the edges that diverge from that vertex. Remember that the area of a spherical triangle with angles α, β, and γ is the "spherical excess" $\alpha + \beta + \gamma - \pi$, and derive hence an expression for the area of a spherical polygon. You obtain so (use ex. 15.5, 15.6)

$$\Sigma\omega = 2\Sigma\delta - \Sigma\pi(n - 2)V_n = 2\Sigma\delta - 2\pi(E - V)$$

15.21. The usual answers are: the equilateral triangle, the square, the regular polygon with n sides.

15.22. Usually, the same shapes are guessed as in ex. 15.21.

15.25. The usual answers are: the regular tetrahedron, the regular octahedron, the cube.

15.26. The usual answers are: the regular tetrahedron, the cube, the regular octahedron.

15.27. A diagonal of the cube is a diameter of the sphere. Hence, if a is the length of an edge of the cube

$$(2r)^2 = 3a^2$$

(*Cf.* HSI, pp. 7–14, the main example of Part I.) Therefore, the required volume is

$$a^3 = \frac{8\sqrt{3}r^3}{9}$$

15.28. Let A denote the area of an equilateral triangle with side r. The required volume is

$$\frac{2 \cdot 6Ar}{3} = \sqrt{3}r^3$$

The answer to the case $V = 8$ of ex. 15.25, which appeared as a quite plausible guess, turned out to be wrong. (The answers to the cases $V = 4$ and 6 are correct, however; for the first, *cf.* MPR, vol. 1, p. 133, ex. 17.)

15.29. Let A be the area and h the altitude of a face of the octahedron. Then the required volume is (divide the octahedron into 8 tetrahedra)

$$V = \frac{8Ar}{3} = \frac{8r}{3} \frac{h^2}{\sqrt{3}}$$

The sphere touches, by symmetry, each face at its center. Hence h is, in a right triangle, the length of the hypotenuse which is divided by the altitude (of length r) into two pieces, of lengths $h/3$ and $2h/3$, respectively. Thus

$$\frac{h}{3} \cdot \frac{2h}{3} = r^2$$

By eliminating h, we find

$$V = 4\sqrt{3}r^3$$

15.30. Volume of the prism is

$$6 \frac{r^2}{\sqrt{3}} 2r = 4\sqrt{3}r^3$$

In answering the case $F = 8$ of ex. 15.26 we hardly expected this. (The answers to cases $F = 4$ and 6 are correct.)

15.31.

	F	V	E
Cube	6	8	12
Hexagonal double pyramid	12	8	18
Regular octahedron	8	6	12
Hexagonal right prism	8	12	18

The nonregular polyhedron that "beats" (or "ties with") the regular polyhedron agrees with it, of course, in the *prescribed* number of elements (V in the first case, F in the second) but is more complicated otherwise (has greater F and E in the first case, greater V and E in the second). Should this circumstance account for the observed failure of the principle of nonsufficient reason?

15.32. One which has only triangular faces, see ex. 15.11.

15.33. One which has only three-edged vertices, see ex. 15.12.

15.34. Between the cube and the octahedron there is a reciprocal relation, and we can observe the same relation between the corresponding rival nonregular polyhedra; *cf.* MPR, vol. 1, p. 53, ex. 3 and 4. This suggests a conjecture: between the *solutions* of the problems of ex. 15.25 and ex. 15.26 for the same number of data the same (topological) *reciprocal* relation will hold; *cf.* ex. 15.32 and 15.33.

15.35. The usual answers are the same as to ex. 15.26, and the whole situation is the same; the usual answer is correct for $F = 4$ and $F = 6$, incorrect for $F = 8$. *Cf.* MPR, vol. 1, p. 188, ex. 42.

15.36. (Stanford 1962.) We are required to find the points of intersection of two congruent ellipses symmetrical to each other with respect to the line $x = y$. Subtraction of the equations yields $x^2 = y^2$. Of the four points of intersection

$$(6, 6) \qquad (-6, -6) \qquad (2, -2) \qquad (-2, 2)$$

two do, and two do not, comply with the principle of nonsufficient reason. *Cf.* ex. 6.22.

15.37. By appropriate subtractions $x^2 = y^2 = z^2$. Of the eight solutions

$$(1, 1, 1) \quad (-1, -1, -1)$$
$$(3, -3, -3) \quad (-3, 3, 3) \quad (-3, 3, -3) \quad (3, -3, 3) \quad (-3, -3, 3) \quad (3, 3, -3)$$

two do, and six do not, comply with the principle of nonsufficient reason.

15.38. (Stanford 1963.) Same solutions as in ex. 15.37, but it is less easy to establish that $x^2 = y^2 = x^2$.

15.41. See ex. 15.42–15.53.

15.42. (1) We can discuss the result of any problem "in letters" in this manner by checking particular cases; *cf.* sect. 2.4(3); ex. 2.61; HSI, Can you check the result? 2, p. 60; MPR, vol. 2, pp. 5–7, sect. 2, pp. 13–14, ex. 3–7; also paper no. 11 of the author, quoted in the Bibliography; etc.

(2) In checking its particular cases, we familiarize ourselves with the formula, we understand its "structure" better. Moreover, such a discussion may illustrate several important general ideas. We may learn that the merit of a formula consists in its generality, in its applicability. Then, we may also learn plausible, inductive reasoning—assessing the likelihood of a general assertion by examining its particular cases. In brief, the teacher who neglects discussions of the kind presented in sect. 15.4 misses a most obvious opportunity to do something for the mental maturity of his students.

15.43. Each point of the region displayed in Fig. 15.1 represents a triangular shape. (For diagrams surveying in an analogous way ellipsoidal and lenticular shapes see G. Pólya and G. Szegö, *Isoperimetric inequalities in mathematical physics,* Princeton University Press, p. 37 and p. 40.) Thus, Fig. 15.1 may prepare the student for a certain way of using diagrams in science, for instance, indicator diagrams in thermodynamics. Moreover Fig. 15.1 offers nonpointless practice in the geometrical interpretation of linear inequalities.

15.44. Here are some of the facts to which experimentation with decimal fractions may lead.

All three types of decimal expansions represent rational numbers and, conversely, the decimal expansion of a rational number must belong to one of the three types. The difference between the types depends on the prime factors of the denominator of the represented rational number; according as all these prime factors divide 10, or none of them divides 10, or there is one prime factor that divides and another that does not divide 10, the decimal fraction is terminating, or purely periodical, or mixed. (In speaking about *the* denominator b of a rational number a/b we have supposed that a/b is in lowest terms, that is, that the integers a and b have no common divisor greater than 1, and $b \geqq 1$. We have disregarded two obvious cases: the case $b = 1$ (integers) and the case of the infinite decimal expansion of such rational numbers as can be represented also by a terminating expansion. *Cf.* I. Niven, *Numbers: rational and irrational,* Random House, pp. 23–26, 30–37.)

The length of the period is independent of the numerator.

If the denominator is a prime number p, the length of the period is a divisor of $p - 1$. (More generally, the length of the period divides $\varphi(b)$, the number of positive integers not exceeding the denominator b and relative prime to b. What can you say about mixed expansions?)

If the denominator is a prime and the length of the period an even number, each digit in the second half of the period completes the corresponding digit in the first half to 9. (For instance, in the expansion

$$\tfrac{1}{7} = 0.\overline{142857}$$

$$1 + 8 = 9 \qquad 4 + 5 = 9 \qquad 2 + 7 = 9$$

The knowledge of this fact may save much work in computing decimal fractions.)

If the denominator is not divisible by 3, the sum of the digits in the period is divisible by 9. For instance

$$\tfrac{15}{41} = 0.\overline{36585}$$

$$3 + 6 + 5 + 8 + 5 = 27$$

The reader should check these statements on examples. The proofs are easy, provided that some knowledge of the theory of numbers is available—to rouse the reader's interest in this theory is one of the aims.

15.45. Observation:

$$9 \times 11 = 99, \quad 27 \times 37 = 999, \quad 99 \times 101 = 9999, \quad 271 \times 369 = 99999$$

Explanation: Therefore, for instance,

$$27 \times 0.037037 \cdots = 0.999999 \cdots = 1$$

We should not hesitate to compare little things with great things (*parva componere magnis*): such a comparison may be instructive. The step we have taken from "observation" to "explanation," from noticing a regularity to noticing the underlying connection, is infinitely smaller in scale than, but similar in nature to, the step from Kepler to Newton; *cf.* MPR, vol. 1, p. 87, ex. 15.

15.46. Except for the last statement (concerning the sum of the digits in the period) to each result of ex. 15.44 there corresponds a parallel result in the binary system. For instance, in the *binary* expansion

$$\tfrac{3}{5} = 0.\overline{1001}$$

the length of the period is $5 - 1$ and

$$1 + 0 = 1, \qquad 0 + 1 = 1$$

15.47. Challenging arithmetical work, practice in decimal fractions and factorization. Broad background: concept of real number ("And what about the decimal expansion of $\sqrt{2}$, or of π?"). A prelude to number theory. On the general cultural level: wide opportunity for inductive reasoning, even for the construction of a comprehensive theory starting from an experimental basis.

Take a detail; ex. 15.45 offers an exceptionally neat opportunity to confirm a guess based on observation by a proof, by the understanding of the underlying connection.

15.48. See ex. 15.49.

15.49. In Fig. 15.11, $t(n)$ is the number of dots with abscissa n. We find that $t(n) = 1$ for 1, 2, 4, 8, 16.

When p is an *odd* prime $t(p) = 2$.

Even after these significant hints (and even after a comparison of Fig. 15.11 with Fig. 15.12) it may take longer experimentation and some acumen to discover the rule: $t(n)$ *equals the number of odd divisors of n.* The reader should prove this rule; he may profit from the following remarks.

(1) The trapezoidal representation displayed in ex. 15.49 is equivalent to the equation

$$2n = r(r + 2a - 1)$$

(2) Of the two factors r and $r + 2a - 1$, one is odd, the other is even, and the odd factor must divide n.

(3) The smaller of these two factors is r, the number of rows.

(4) If n and r are given, a is uniquely determined.

15.50. We use the symbol $\tau(n)$, as defined in ex. 9.12. The rule distinguishes five cases:

(1) If n is odd and not a square, $s(n) = \tau(n)/2$.
(2) If n is odd and a square, $s(n) = [\tau(n) + 1]/2$.
(3) If n is even and not divisible by 4, $s(n) = 0$.
(4) If n is divisible by 4 and not a square, $s(n) = \tau(n/4)/2$.
(5) If n is divisible by 4 and a square, $s(n) = [\tau(n/4) + 1]/2$.

To prove this rule, observe that

$$n = (2a + 1) + (2a + 3) + \cdots + (2a + 2r - 1) = r(r + 2a)$$

If n is divisible by 4, observe that

$$\frac{n}{4} = \frac{r}{2}\left(\frac{r}{2} + a\right)$$

15.51. Let us compare the present project with that assessed in ex. 15.47. In the present case the problem is more artificial, the background less rich, the law more difficult to guess, but the proof, although challenging, needs very little preliminary knowledge; I think that the project is very much worthwhile.

Figure 15.11 yields a nontrivial, instructive example of a binary relation (between the two positive integers n and r; n is the sum of r successive positive integers) and of its representation by a diagram. For Fig. 15.12, which represents a better known and more important binary relation, see Leibnitz, *Opuscules*, p. 580. The study of these diagrams could profitably precede the introduction of the concept "binary relation."

15.52. The required quantities can be computed without explicit use of the integral calculus (Cavalieri's principle, Pappus' rules). They are displayed in the table:

	Double Cone	Sphere	Cylinder	
V	$\dfrac{2\pi a^3}{3}$	$\dfrac{4\pi a^3}{3}$	$2\pi a^3$	$1:2:3$
S	$2\sqrt{2}\,\pi a^2$	$4\pi a^2$	$6\pi a^2$	$\sqrt{2}:2:3$
A	$2a^2$	πa^2	$4a^2$	$2:\pi:4$
L	$4\sqrt{2}\,a$	$2\pi a$	$8a$	$2\sqrt{2}:\pi:4$
X_A	$\dfrac{a}{3}$	$\dfrac{4}{3\pi}a$	$\dfrac{a}{2}$	$\dfrac{X_A}{X_L}=\dfrac{2}{3}$
X_L	$\dfrac{a}{2}$	$\dfrac{2}{\pi}a$	$\dfrac{3}{4}a$	

$$S = \sqrt{2}\,\frac{dV}{da} \qquad S = \frac{dV}{da} \qquad S = \frac{dV}{da}$$

$$L = \sqrt{2}\,\frac{dA}{da} \qquad L = \frac{dA}{da} \qquad L = \frac{dA}{da}$$

For a generalization of the observation about X_A/X_L (which, by the way, turned up in a class discussion) see a paper by C. J. Gerriets and the author, *American Mathematical Monthly,* vol. 66, 1959, pp. 875–879.

15.53. For $n = 1, 2, 3, \ldots$

$$(\sqrt{2} - 1)^n = \sqrt{m + 1} - \sqrt{m}$$

where m is a positive integer which depends on n. Proof by (easy) mathematical induction. See *American Mathematical Monthly,* vol. 58, 1951, p. 566.

No solution: **15.23, 15.24, 15.39, 15.40, 15.54, 15.55.**

BIBLIOGRAPHY

I. Classics

EUCLID, *Elements*. The inexpensive shortened edition in Everyman's Library is sufficient here. ("Euclid III 20" refers to Proposition 20 of Book III of the Elements.)

PAPPUS ALEXANDRINUS, *Collectio*, edited by F. Hultsch, 1877; see v. 2, pp. 634–637 (the beginning of Book VII).

DESCARTES, *Œuvres*, edited by Charles Adam and Paul Tannery. (For remarks on the "Rules"—the work which is of especial interest for us—and the way of quoting it, see ex. 2.72.)

LEIBNITZ (or LEIBNIZ) (1) *Mathematische Schriften*, edited by C. J. Gerhardt. (2) *Philosophische Schriften*, edited by C. J. Gerhardt. (3) *Opuscules et fragments inédits*, collected by Louis Couturat.

BERNARD BOLZANO, *Wissenschaftslehre*, second edition, 1930; see v. 3, pp. 293–575 (Erfindungskunst).

II. More modern

E. MACH, *Erkenntnis und Irrtum*, fourth edition, Leipzig, 1924; see pp. 251–274 and *passim*.

J. HADAMARD, *Leçons de Géométrie plane*, Paris, 1898; see Note A, *Sur la méthode en géométrie*.

F. KRAUSS, Denkform mathematischer Beweisführung, *Zeitschrift für mathematischen und naturwissenschaftlichen Unterricht*, v. 63, pp. 209–222.

WERNER HARTKOPF, *Die Strukturformen der Probleme;* Dissertation, Berlin, 1958.

I. LAKATOS, Proofs and Refutations, *The British Journal for the Philosophy of Science*, v. 14, 1963, pp. 1–25, 120–139, 221–245.

FRANZ DENK, WERNER HARTKOPF, and GEORGE POLYA, Heuristik, *Der Mathematikunterricht*, v. 10, 1964, part 1.

III. Related work of the author

Books

1. *Aufgaben und Lehrsätze aus der Analysis*, 2 volumes; second unchanged edition, Berlin, 1954; jointly with G. SZEGÖ.

184

2. *How to Solve It;* second edition, 1957; Anchor Book A 93, Doubleday. (Quoted as HSI; references are to the pages of the second edition but, for the convenience of the users of former printings, titles are added.)

3. *Mathematics and Plausible Reasoning,* Princeton, 1954. In two volumes, entitled *Induction and Analogy in Mathematics* (vol. I) and *Patterns of Plausible Inference* (vol. II). (Quoted as MPR.)

4. *Mathematical Methods in Science,* lectures edited by *Leon Bowden.* School Mathematics Study Group Studies in Mathematics, v. XI.

Papers

1. Geometrische Darstellung einer Gedankenkette. *Schweizerische Pädagogische Zeitschrift,* 1919, 11 pp.

2. Wie sucht man die Lösung mathematischer Aufgaben? *Zeitschrift für mathematischen und naturwissenschaftlichen Unterricht,* **63,** 1932, pp. 159–169.

3. Wie sucht man die Lösung mathematischer Aufgaben? *Acta Psychologica,* **4,** 1938, pp. 113–170.

11. Die Mathematik als Schule des plausiblen Schliessens. *Gymnasium Helveticum,* **10,** 1956, pp. 4–8; reprinted *Archimedes,* **8,** 1956, pp. 111–114. Mathematics as a subject for learning plausible reasoning; translation by C. M. Larsen. *The Mathematics Teacher,* **52,** 1959, pp. 7–9.

12. On picture-writing. *American Mathematical Monthly,* **63,** 1956, pp. 689–697.

13. L'Heuristique est-elle un sujet d'étude raisonnable? *La Méthode dans les Sciences Modernes* ("Travail et Méthode," numéro hors série) 1958, pp. 279–285.

14. On the curriculum for prospective high school teachers. *American Mathematical Monthly,* **65,** 1958, pp. 101–104.

15. Ten Commandments for Teachers. *Journal of Education of the Faculty and College of Education, Vancouver and Victoria,* nr. 3, 1959, pp. 61–69.

16. Heuristic reasoning in the theory of numbers. *American Mathematical Monthly,* **66,** 1959, pp. 375–384.

17. Teaching of Mathematics in Switzerland. *American Mathematical Monthly,* **67,** 1960, pp. 907–914; *The Mathematics Teacher,* **53,** 1960, pp. 552–558.

18. The minimum fraction of the popular vote that can elect the President of the United States. *The Mathematics Teacher,* **54,** 1961, pp. 130–133.

19. The Teaching of Mathematics and the Biogenetic Law. *The Scientist Speculates,* edited by *I. J. Good,* 1962, pp. 352–356.

20. On Learning, Teaching, and Learning Teaching. *American Mathematical Monthly,* **70,** 1963, pp. 605–619.

(The numbers omitted are quoted in MPR, v. 1, pp. 279–280, and v. 2, pp. 189–190.)

IV. Problems

Among the examples proposed for solution, some are taken from the *Stanford University* (*Stanford-Sylvania*) *Competitive Examination in Mathematics.* This fact is indicated at the beginning of the solution with the year in which the problem was given as "Stanford 1957." Most of these problems (some with solution) appeared in *The California Mathematics Council Bulletin.*

The USSR Olympiad Problem Book, by *Shklarsky, Chentzov,* and *Yaglom,* contains many unusual and difficult elementary problems proposed in Russian competitive examinations. English translation, edited by *I. Sussman,* San Francisco, 1962.

Hungarian Problem Book, New Mathematical Library, v. 11–12, contains interesting elementary problems, careful solutions, instructive notes on the subject matters involved, and remarks on a competitive examination which essentially contributed to the development of mathematics in Hungary.

V. On Teaching

On the Mathematics Curriculum of the High School (memorandum with 65 signatures). *American Mathematical Monthly,* **69,** 1962, pp. 189–193; *The Mathematics Teacher,* **55,** 1962, pp. 191–195.

MARTIN WAGENSCHEIN, Exemplarisches Lehren im Mathematikunterricht. *Der Mathematikunterricht,* **8,** 1962, part 4.

A. I. WITTENBERG, *Bildung und Mathematik,* Stuttgart, 1963.

A. I. WITTENBERG, SOEUR SAINTE-JEANNE-DE-FRANCE, and F. LEMAY, *Redécouvrir les mathématiques,* Neuchâtel, 1963.

ROY DUBISCH, *The Teaching of Mathematics,* New York, 1963. Contains also extensive annotated references to recent literature in English.

APPENDIX

Problems

Problems 1.19.1, 1.19.2, and 1.19.3 approximately follow (in this order) ex. 1.19, problem 2.27.1 follows ex. 2.27, etc.

1.19.1. Triangle from α, r, R.
[*Could you think of other data more appropriate to determine the unknown? Could you exchange just one datum for a more suitable one?*]

1.19.2. Triangle from a, h_a, r.
[*Could you derive something useful from the data?*]

1.19.3. Triangle from a, r, $a + b + c$.

2.27.1. *How long did Diophantus live?* The problem is proposed in the form of an alleged inscription on Diophantus' tombstone. The original is in verse. (*Cf.* B. L. van der Waerden, *Science Awakening*, p. 278.)
Here is the tomb of Diophantus. This stone tells you his age if you have mastered his art. The gods gave him to live one sixth of his life as a boy. After one twelfth more his beard began to show. Then one seventh of his life passed till his wedding day. After five years of marriage a boy was born. Alas, his beloved son came to an untimely end attaining only one half of the age he himself was permitted to live. After this bereavement he sought consolation in mathematics for four years and then he too ended his earthly career.

2.35.1. From a vertex of a certain triangle draw the altitude, the angle bisector, and the median. Being given that these three lines divide the angle α at the vertex into four equal parts, find α.
[You may wish to know also the shape of the triangle. Pay attention to *each part of the condition.*]

2.40.1. Of a right triangle, given the length of the hypotenuse c and the area A. On each side of the triangle, describe a square exterior to the triangle and consider the least convex figure containing the three squares (formed by a tight rubber band around them): it is a hexagon (it is irregular, has one side

187

in common with each square, and one of its remaining three sides is obviously of length c).

Find the area of the hexagon.

2.40.2. In a right triangle, c is the length of the hypotenuse, a and b are the lengths of the two other sides, and d is the length of the diameter of the inscribed circle. Prove that

$$a + b = c + d$$

[*Restate the problem:* Given a, b, and c, compute d.]

2.50.1. Here is a problem of solid geometry analogous to ex. 2.45.

Start by dissecting the three-dimensional space into equal cubes.

First pattern: Each cube is associated with a concentric sphere that touches its six faces.

Second pattern: "Each second" cube is associated with a concentric sphere that touches its 12 edges. (That is, of two cubes that have a face in common, one does and the other does not contain the center of an associated sphere.)

Compute the percentage of space contained in the spheres for each pattern.

2.52.1. A cake has the shape of a right prism with a square base; it has icing on the top as well as on the sides (that is, on the four lateral faces). The altitude of the prism is $\frac{5}{16}$ of the side of its base. Cut the cake into 9 pieces so that each piece has the same amount of cake and the same amount of icing. One of the 9 pieces should be a right prism with a square base with icing only on the top: Compute the ratio of its altitude to a side of its base and give a clear description of all 9 pieces.

2.55.1. Five edges of a tetrahedron are of the same length a, and the sixth edge is of the length b.

(1) Express the radius of the sphere circumscribed about the tetrahedron in terms of a and b.

(2) How would you use the result (1) to determine practically the radius of a spherical surface (of a lens)?

2.55.2. *The carbon atom* has four valencies which we picture as directed symmetrically in space. From the center of a regular tetrahedron draw four lines to the four vertices and compute the angle α included by any two of them.

2.55.3. *Photometer.* A lamp L has the candle-power I, another lamp L' the candle-power I', and their distance is d. Find the position of a screen that is so placed between these lamps, perpendicularly to the line joining them, that it is equally illuminated from both sides.

(If the candle-power of a point-source L of light is I, the illumination of a surface, which is at the distance x from L and is perpendicular to the distance, is I/x^2. To understand this thoroughly, you should consider two spheres with center L: one with radius 1, the other with radius x.)

3.10.1. *Salvaging.* The ship sank—perhaps there is some treasure in it worth lifting. Your plan has failed—perhaps there is an idea in it worth saving.

In sect. 3.2, our first plan to compute S_2 (notation of sect. 3.3) failed ignominiously: The procedure that worked for S_1 did not work at all as we tried to apply it to S_2. Where does the fault lie? Perhaps we applied the procedure too rigidly. What about applying it less rigidly? Applying it with some modification? Or applying it to some other case?

Such consideration may lead to trials, and it is quite natural to try the procedure on S_k. What is the essential point of the procedure? Combining two terms equally removed from the ends: one term as distant from one end as the other is from the other end. Such are the terms j^k and $(n - j)^k$ of S_k. If addition does not work, subtraction may—and eventually we may adapt the combination to the nature of k:

$$(n - j)^k - (-j)^k = n^k - \binom{k}{1}n^{k-1}j + \binom{k}{2}n^{k-2}j^2 - \cdots + (-1)^{k-1}\binom{k}{k-1}nj^{k-1}$$

Write this successively for $j = 0, 1, 2, \ldots, n - 1, n$:

$$n^k - (-1)^k 0^k = n^k$$

$$(n - 1)^k - (-1)^k 1^k = n^k - \binom{k}{1}n^{k-1}1 + \binom{k}{2}n^{k-2}1^2$$
$$- \cdots + (-1)^{k-1}\binom{k}{k-1}n1^{k-1}$$

$$(n - 2)^k - (-1)^k 2^k = n^k - \binom{k}{1}n^{k-1}2 + \binom{k}{2}n^{k-2}2^2$$
$$- \cdots + (-1)^{k-1}\binom{k}{k-1}n2^{k-1}$$

. . . .

$$1^k - (-1)^k(n - 1)^k = n^k - \binom{k}{1}n^{k-1}(n - 1) + \binom{k}{2}n^{k-2}(n - 1)^2$$
$$- \cdots + (-1)^{k-1}\binom{k}{k-1}n(n - 1)^{k-1}$$

$$0^k - (-1)^k n^k = n^k - \binom{k}{1}n^{k-1}n + \binom{k}{2}n^{k-2}n^2$$
$$- \cdots + (-1)^{k-1}\binom{k}{k-1}nn^{k-1}$$

By adding and using the notation of sect. 3.3 (but writing S_0' for $S_0 + 1$) we obtain

$$S_k[1 - (-1)^k] = n^k S_0' - \binom{k}{1}n^{k-1}S_1 + \binom{k}{2}n^{k-2}S_2$$
$$- \cdots + (-1)^{k-1}\binom{k}{k-1}nS_{k-1}$$

Explore the cases $k = 1, 2, 3$ of this result and then try to assess it generally.

3.40.1. In how many ways can the positive integer n be the sum of positive integers? In how many ways can n be the sum of a specified number t of terms (all positive integers)? Let us regard two sums that differ in the order of terms as different.

It is natural to start exploring such a problem by experiments and by trying to arrange neatly the material produced by the experiments. Here are the sums found for $n = 4$ and $n = 5$:

4	1 + 3	2 + 1 + 1	1 + 1 + 1 + 1
	2 + 2	1 + 2 + 1	
	3 + 1	1 + 1 + 2	

5	1 + 4	3 + 1 + 1	2 + 1 + 1 + 1	1 + 1 + 1 + 1 + 1
	2 + 3	1 + 3 + 1	1 + 2 + 1 + 1	
	3 + 2	1 + 1 + 3	1 + 1 + 2 + 1	
	4 + 1	1 + 2 + 2	1 + 1 + 1 + 2	
		2 + 1 + 2		
		2 + 2 + 1		

Do you notice a pattern?
Prove your guess!
Could a geometric figure help you?

3.40.2. *Fibonacci numbers.* By adding the numbers along the oblique lines of Fig. A3.40.2 we obtain the sequence of the Fibonacci numbers

$$1, 1, 2, 3, 5, 8, 13, 21, \ldots$$

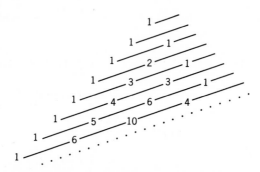

Fig. A3.40.2. An oblique approach to Fibonacci.

We let F_n stand for its nth term, the nth Fibonacci number; thus $F_1 = 1$, $F_2 = 1$, $F_8 = 21$.

(1) Express F_n in terms of binomial coefficients.

(2) Prove that for $n = 3, 4, 5, \ldots$

$$F_n = F_{n-1} + F_{n-2}$$

3.40.3. (*Continued*). The sequence of numbers

$$1, 1, 1, 2, 3, 4, 6, 9, 13, \ldots$$

is analogously generated (see Fig. A3.40.3). Let G_n stand for its nth term; thus $G_9 = 13$.

(1) Express G_n in terms of binomial coefficients.

(2) Prove that for $n = 4, 5, 6, \ldots$

$$G_n = G_{n-1} + G_{n-3}$$

(3) Generalize.

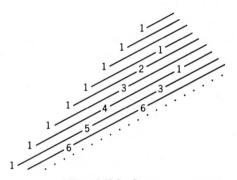

Fig. A3.40.3. Steeper.

†**3.60.1.** Consider the table

			$1 \cdot 1$				$= 1$
		$1 \cdot 3 -$	$2 \cdot 2 +$	$3 \cdot 1$			$= 2$
	$1 \cdot 5 -$	$2 \cdot 4 +$	$3 \cdot 3 -$	$4 \cdot 2 +$	$5 \cdot 1$		$= 3$
$1 \cdot 7 -$	$2 \cdot 6 +$	$3 \cdot 5 -$	$4 \cdot 4 +$	$5 \cdot 3 -$	$6 \cdot 2 +$	$7 \cdot 1 = 4$

Guess the general law suggested by these examples, express it in suitable mathematical notation, and prove it.

3.65.1. If x and n stand for positive integers the expression

$$\frac{x^2(x^2 - 1)(x^2 - 4) \cdots [x^2 - (n - 1)^2]}{(2n - 1)!n}$$

represents an integer.

†**3.88.1.** Another solution of ex. 3.86, which is simpler in some respects, uses differential calculus. Find it.

†**3.88.2.** Observe that

$$
\begin{aligned}
1 \cdot 1 + 2 \cdot 1 &= 3 \\
1 \cdot 1 + 2 \cdot 2 + 3 \cdot 1 &= 8 \\
1 \cdot 1 + 2 \cdot 3 + 3 \cdot 3 + 4 \cdot 1 &= 20 \\
1 \cdot 1 + 2 \cdot 4 + 3 \cdot 6 + 4 \cdot 4 + 5 \cdot 1 &= 48 \\
1 \cdot 1 + 2 \cdot 5 + 3 \cdot 10 + 4 \cdot 10 + 5 \cdot 5 + 6 \cdot 1 &= 112
\end{aligned}
$$

Guess the general law suggested by these examples, express it in suitable mathematical notation, and prove it.

4.15.1. Define y_k for $k = 2, 3, 4, \ldots$ by the recursion formula

$$y_k = \frac{y_{k-1} + y_{k-2}}{2}$$

and set

$$y_0 = a, \qquad y_1 = b$$

Express y_k in terms of a, b, and k.

5.19.1. *Examining the solution* of ex. 5.19, we may observe that the number found has a simple interpretation: It is the number of ways of choosing v boxes out of $n + v$ boxes. We should be able to see such a simple thing by a simple argument.

Imagine the $n + v$ boxes in a row—each occupying, if you wish, a subinterval of length 1 of the interval $0 \leqq x \leqq n + v$. What has the proposed problem to do with marking v out of these $n + v$ boxes?

6.13.1. Prove that no number in the sequence

$$11, 111, 1111, 11111, \ldots$$

is the square of an integer.

6.17.1. "How many children have you, and how old are they?" asked the guest, a mathematics teacher.

"I have three boys," said Mr. Smith. "The product of their ages is 72 and the sum of their ages is the street number."

The guest went to look at the entrance, came back and said, "The problem is indeterminate."

"Yes, that is so," said Mr. Smith, "but I still hope that the oldest boy will some day win the Stanford competition."

Tell the ages of the boys, stating your reasons.

7.2.1. There is an interpretation of the diagram in Fig. A7.2.1. which is of historic interest. Can you recognize it?

†9.11.1. *Any solution will do.* Show that there exists a couple of divergent series.

$$a_1 + a_2 + \cdots + a_n + \cdots \qquad b_1 + b_2 + \cdots + b_n + \cdots$$

with positive decreasing terms

$$a_1 > a_2 > a_3 > \cdots \qquad b_1 > b_2 > b_3 > \cdots$$

such that the series

$$\min(a_1, b_1) + \min(a_2, b_2) + \cdots + \min(a_n, b_n) + \cdots$$

is convergent. As usual, $\min(a, b)$ denotes the lesser (not greater) of the two numbers a and b.

[Not all couples of series satisfying the condition stated are required, but just one such couple (any one). Thus Leibnitz's advice quoted in ex. 9.11 is applicable: Narrow down the region of search without increasing your difficulty.]

Fig. A7.2.1. Where?

12.2.1. Prove the proposition: If a side of a triangle is less than the average (arithmetic mean) of the two other sides, the opposite angle is less than the average of the two other angles.

(What are the principal parts? *Express them in mathematical language,* by using the usual symbols of trigonometry.)

12.5.1. *Relevant knowledge.* A quadrilateral with sides a,b,c,d and area A is both inscribed and circumscribed (inscribed in a circle and circumscribed about another circle). Then

$$A^2 = abcd$$

[To prove this proposition may be easy or difficult according as you do or do not know a certain related proposition.]

12.9.1. *Do you know a related problem?* Solve the following system of three equations for the unknowns $x, y,$ and z ($a, b,$ and c are given):

$$x^2y^2 + x^2z^2 = axyz$$
$$y^2z^2 + y^2x^2 = bxyz$$
$$z^2x^2 + z^2y^2 = cxyz$$

(We have here a system of 3 equations with 3 unknowns. The best known systems of this kind are linear: Can we "linearize" the proposed system? We may hit upon the following form:

$$1/y^2 + 1/z^2 = \frac{a}{xyz}$$

$$1/x^2 \qquad + 1/z^2 = \frac{b}{xyz}$$

$$1/x^2 + 1/y^2 \qquad = \frac{c}{xyz}$$

which is linear in x^{-2}, y^{-2}, z^{-2} *if* we regard (wishful thinking) xyz as known. The solution will be of the form

$$\frac{1}{x^2} = \frac{A}{xyz} \cdots$$

and opens a new perspective—can you see it?)

12.9.2. *Go back to definitions.* We consider three circles f, f', and v and their centers F, F', and V, respectively. The circles f and f' are fixed, v is variable, f' and v are inside f but outside each other. Prove the proposition: If the variable circle v touches both fixed circles f and f', the locus of its center V is an ellipse. [What is an ellipse?]

12.9.3. *Exploring the neighborhood.* Did you like the foregoing problem (ex. 12.9.2)? Did you like its solution? Then examine its neighborhood—you found a nice ripe apple on this tree, there may be more.

Vary the problem: You may consider a generalization, or particular cases, limiting cases, analogous cases. There is a chance to find something interesting as well as a chance to learn to do research.

The reader should find the various loci for V that result from the following modifications of the assumptions concerning the fixed circles f and f' and the variable circle v.

(1) *Specialization.* The circles f and f' are concentric.

(2) *Limiting case.* Let f be a fixed straight line and f' a fixed point; v touches f and passes through $f' = F'$.

(3) *Analogy.* The two circles f and f' are outside each other and v touches f and f' in the same way, it is either outside both or it contains both.

(4) Limiting case of (3). Let f and f' be two distinct points and let v pass through both.

Consider other particular, limiting, or analogous cases.

14.1.1. *Leap years.* An ordinary year has 365 days, a leap year 366 days. The year n which is not a centenary year is a leap year if, and only if, n is a multiple of 4. A centenary year, that is, a year n where n is a multiple of 100, is a leap year if (and only if) n is a multiple of 400. Thus 1968 and 2000 are leap years, 1967 and 1900 are not. These rules were established by the Pope Gregory XIII.

So far we spoke of "civil" years which must consist of a whole number of days. The astronomical year is the period of time in which the earth completes a full revolution around the sun. If the Gregorian rules would precisely agree with it, what would be the length of the astronomical year?

15.2.1. [Sect. 15.5.] Let a,b, and c denote the lengths of the sides of a triangle and d the length of the bisector of the angle opposite to the side of length c, terminated on the side.

(1) Express d in terms of a,b, and c.

(2) Check the expression obtained in the four cases illustrated by Figs. 15.2 to 15.5.

15.52.1. *One more high-school level research project,* which could also be proposed as part of a term paper in a class for teachers.

A point (x,y,z) in three-dimensional space is characterized in the usual way by its three rectangular coordinates x,y, and z.

We consider four sets of points, C,O,I, and H. Each set is characterized by a system of inequalities (which may consist of just one inequality): Those points (and only those points) belong to the set the coordinates of which satisfy simultaneously all inequalities of the system.

We list the four systems of inequalities defining our sets:

(C) $|x| \leq 1$ $|y| \leq 1$ $|z| \leq 1$
(O) $|x| + |y| + |z| \leq 2$
(I)   all four inequalities listed under (C) and (O)
(H) $|y| + |z| \leq 2$ $|z| + |x| \leq 2$ $|x| + |y| \leq 2$

Describe carefully the geometric nature of the four sets, mention all relevant features (do not forget symmetries) arranging them clearly in suitable tables.

Describe also the relations between the four figures.

Find the volume V and the surface area S of each figure.

Which generalizations does this work suggest to you?

(Cardboard models may be helpful. *Cf.* ex. 2.50.1, HSI p. 235, ex. 8.)

15.53.1. Observe also that

$$2 - \sqrt{3} = \sqrt{4} - \sqrt{3}$$
$$(2 - \sqrt{3})^2 = \sqrt{49} - \sqrt{48}$$
$$(2 - \sqrt{3})^3 = \sqrt{676} - \sqrt{675}$$
$$(2 - \sqrt{3})^4 = \sqrt{9409} - \sqrt{9408}$$

try to generalize and prove your guess.

Solutions

1.19.1. From the center of the circumscribed circle, draw a line to one of the endpoints of the side a and a perpendicular to a. You thus obtain a right triangle with hypotenuse R, angle α, and opposite leg $a/2$. In our case R and α are given and so you can construct a. Construct the desired triangle from the a so found and the α and r originally given by using ex. 1.19.

1.19.2. Divide the desired triangle into three triangles by joining the center of the inscribed circle to the three vertices. Consideration of the areas yields

$$\tfrac{1}{2} r(a + b + c) = \tfrac{1}{2} a h_a$$

Thus you can construct the length of the perimeter $a + b + c$ from the given a, h_a, r and so reduce the proposed problem to ex. 1.19.3.

1.19.3. From the center of the inscribed circle, draw a line to the vertex A and a perpendicular to the side b (or c). You thus obtain a right triangle in which one of the acute angles is $\alpha/2$ and the opposite leg r. Let x denote the other leg. Now

$$a + b + c - 2a = 2x$$

and so you can first construct x from the given $a + b + c$ and a, and then construct α from x and the given r. Construct the desired triangle from the α so found and a and r originally given by using ex. 1.19.

2.27.1. Let x stand for the number of years Diophantus lived. From

$$\frac{x}{6} + \frac{x}{12} + \frac{x}{7} + 5 + \frac{x}{2} + 4 = x$$

we find $x = 84$.

2.35.1. (*Cf. American Mathematical Monthly*, vol. 66, 1959, p. 208.) Let β be the larger and γ the smaller of the two remaining angles. If β is an acute angle, the five lines c, h_a, d_α, m_a, b starting from the vertex A (notation ex. 1.7) follow each other in this order. From the right triangles into which h_a divides the triangle considered

$$\beta = \frac{\pi}{2} - \frac{\alpha}{4}, \qquad \gamma = \frac{\pi}{2} - \frac{3\alpha}{4}$$

From the triangles into which m_a divides the triangle considered

$$\frac{a/2}{m_a} = \frac{\sin\dfrac{\alpha}{4}}{\sin\left(\dfrac{\pi}{2} - \dfrac{3\alpha}{4}\right)} = \frac{\sin\dfrac{3\alpha}{4}}{\sin\left(\dfrac{\pi}{2} - \dfrac{\alpha}{4}\right)}$$

and so

$$\sin\frac{\alpha}{4} \cos\frac{\alpha}{4} = \sin\frac{3\alpha}{4} \cos\frac{3\alpha}{4}$$

$$\sin\frac{\alpha}{2} = \sin\frac{3\alpha}{2}$$

$$\frac{\alpha}{2} = \pi - \frac{3\alpha}{2}$$

$$\alpha = \frac{\pi}{2}$$

2.40.1. (Stanford 1965.) The hexagon consists of three squares and four triangles and these triangles have all the same area A. Therefore the hexagon's area is $2c^2 + 4A$.

2.40.2. (Stanford 1963.) Divide the given right triangle into three triangles with a common vertex at the center of the circle inscribed in the given. From comparing the areas

$$\frac{d}{2}\frac{a+b+c}{2}=\frac{ab}{2}$$

$$d = \frac{2ab}{a+b+c} = \frac{2ab(a+b-c)}{(a+b)^2 - c^2} = a+b-c$$

2.50.1. $100\pi/6$ and $100\pi\sqrt{2}/6$ or approximately 52.36% and 74.07%, respectively. *Cf.* the solution of ex. 15.52.1, part (6).

2.52.1. (Stanford 1964.) Let C stand for the given prism (the cake) and D for the desired prism (with icing only on the top). For a side of the base and the altitude, respectively, let

$$s \text{ and } h \text{ in } C$$
$$x \text{ and } y \text{ in } D$$

stand. The conditions defining D are expressed by the equations

$$x^2 = \frac{s^2 + 4sh}{9}$$

$$x^2 y = \frac{s^2 h}{9}$$

$$h = \frac{5s}{16}$$

which yield

$$x = \frac{s}{2} \qquad y = \frac{5s}{36} \qquad \frac{y}{x} = \frac{5}{18}$$

Carve out D from C so that either the sides or the diagonals of its top are parallel to the sides of the top of C, and in either case so that the prisms C and D have the same 4 planes of symmetry, which cut the remainder of C into 8 congruent pieces each having the same volume and the same amount of "icing" as D.

2.55.1. (Stanford 1962.) Let C denote the center and r the radius of the circumsphere. There are two "key plane figures," two cross sections of the tetrahedron, one through the edge b and the midpoint of the opposite edge, the other through this latter edge and the midpoint of the edge b. These two cross sections are perpendicular to each other; their intersection d joins the two midpoints considered and contains C.

Let x denote the perpendicular distance of C from b (its other end is the midpoint of b) and h the altitude of one of the two faces that are equilateral triangles; therefore

$$h^2 = \frac{3a^2}{4}$$

From right triangles in the two cross sections we find the equations

$$h^2 = d^2 + \left(\frac{b}{2}\right)^2$$

$$r^2 = x^2 + \left(\frac{b}{2}\right)^2$$

$$r^2 = (d - x)^2 + \left(\frac{a}{2}\right)^2$$

and now we have four equations to determine our four unknowns: h is immediately obtained from the first, then d from the second. It is convenient to find x after d (subtract from each other the last two equations). Finally,

$$r^2 = \frac{a^2}{4} \frac{4a^2 - b^2}{3a^2 - b^2}$$

(*Check:* If $b = a\sqrt{3} = 2h$, then $r = \infty$.)

Possible application: Two rigid equilateral triangles with side a and a common side as hinge can be so inclined to each other that all four vertices are on a concave spherical surface; then by measuring b we obtain r. A convex lens demands a somewhat more sophisticated gadget.

2.55.2. If in ex. 2.55.1 we take $a = b$, the tetrahedron becomes regular,

$$r^2 = \frac{3a^2}{8}$$

and

$$\sin \frac{\alpha}{2} = \frac{a/2}{r} = \frac{\sqrt{6}}{3}$$

$$\alpha = 109°28'$$

This may be supposed to be the angle between any two valences of a carbon atom symmetrically bound (as the one in CH_4).

2.55.3. Let x be the perpendicular distance from L to the screen. Then

$$\frac{I}{x^2} = \frac{I'}{(d - x)^2}$$

and so

$$x = \frac{d\sqrt{I}}{\sqrt{I} + \sqrt{I'}}$$

(The practical problem is somewhat different: We are given I, we measure d and x; hence we have to determine I'.)

3.10.1. Here are the first three particular cases:

$$2S_1 = n(n + 1)$$
$$0 = n^2(n + 1) - 2nS_1$$
$$2S_3 = n^3(n + 1) - 3n^2S_1 + 3nS_2$$

The case $k = 1$ yields the evaluation of S_1 by a method which differs but little from the "method of little Gauss" (sect. 3.1).

The case $k = 2$ leads, by a roundabout way, again to S_1.

The case $k = 3$ yields S_3 provided that S_1 and S_2 are already known.

In general, the result allows to compute S_k from the foregoing S_0', S_1, $S_2, \ldots, S_{k-1}$ *when k is odd,* but not when k is even. This explains to some extent (by the generalization of a modification) why the method that succeeded for S_1 in sect. 3.1 failed for S_2 in sect. 3.2. Morever, by comparing our result with sects. 3.2 to 3.4 and exs. 3.6 to 3.10, we may learn a little—and somebody could use it someday for some purpose.

3.40.1. I hope the reader has explored also the cases $n = 1,2,3$.

Guess: There are $\binom{n-1}{t-1}$ different ways of representing n as the sum of t positive integers.

The cases $t = 1, n$ are trivial, the cases $t = 2, n - 1$ fairly obvious. For a general proof, consider on the number line the interval $0 \leq x \leq n$ and on it the points $x = 1,2,3,\ldots,n - 1$. In choosing any $t - 1$ out of these $n - 1$ points as points of division, we cut the interval into t successive subintervals of integral length and so we make n a sum of t successive terms of the desired kind.

For teachers: The Cuisenaire rods can attractively present this problem and its solution.

3.40.2. Check relations on Fig. A3.40.2 for what values of n you can.

(1) $F_n = \binom{n-1}{0} + \binom{n-2}{1} + \binom{n-3}{2} + \cdots$

(2) From the recursion formula, see sect. 3.6(2). *Cf.* ex. 4.15.

3.40.3.

(1) $G_n = \binom{n-1}{0} + \binom{n-3}{1} + \binom{n-5}{2} + \cdots$

(2) From the recursion formula.

(3) Changing the slope leads to the sequence $y_1, y_2, y_3, \ldots$, dependent on one parameter (the slope, or the positive integer q) satisfying the recursion formula (difference equation, ex. 4.14)

$$y_n = y_{n-1} + y_{n-q}$$

In the case $q = 1$ the slope is 0 and

$$y_n = 2y_{n-1}$$

See ex. 3.32.

3.60.1. Guess:

$$1(2n - 1) - 2(2n - 2) + 3(2n - 3) - \cdots + (2n - 1)1 = n$$

To prove this, consider the coefficient of x^{2n-2} in the product

$$(1 + 2x + 3x^2 + 4x^3 + \cdots)(1 - 2x + 3x^2 - 4x^3 + \cdots)$$
$$= (1 - x)^{-2}(1 + x)^{-2}$$
$$= (1 - x^2)^{-2}$$
$$= 1 + 2x^2 + 3x^4 + 4x^6 + \cdots + nx^{2n-2} + \cdots$$

3.65.1. (*Cf.* Stanford 1963.) Observe that

$$\frac{x}{n} = \frac{(x + n) + (x - n)}{2n}$$

Hence the proposed expression equals

$$\binom{x+n}{2n} + \binom{x+n-1}{2n}$$

and binomial coefficients are integers.

3.88.1. Differentiate both sides of

$$1 + x + x^2 + \cdots + x^n = \frac{1 - x^{n+1}}{1 - x}$$

Ex. 3.87 and 3.88 can be similarly treated.

3.88.2. Guess:

$$1\binom{n}{0} + 2\binom{n}{1} + 3\binom{n}{2} + \cdots + (n+1)\binom{n}{n} = (n+2)2^{n-1}$$

(The difficulty in arriving at this guess may have been recognizing the products $3 \cdot 1, 4 \cdot 2, 5 \cdot 4, 6 \cdot 8, 7 \cdot 16$.)

For a proof, first differentiate both sides of the identity

$$\binom{n}{0}x + \binom{n}{1}x^2 + \binom{n}{2}x^3 + \cdots + \binom{n}{n}x^{n+1} = x(1 + x)^n$$

and then set $x = 1$.

4.15.1. The equation in r

$$2r^2 - r - 1 = 0$$

has the roots $r = 1$ and $r = -1/2$. Therefore, by ex. 4.14,

$$y_k = c_1 + \frac{(-1)^k c_2}{2^k}$$

By using the initial conditions (the cases $k = 0$ and $k = 1$), we obtain c_1, c_2, and, finally,

$$y_k = \frac{a + 2b}{3} + (-1)^k \frac{a - b}{3 \cdot 2^{k-1}}$$

5.19.1. We put a mark (a multiplication point if you wish) on v out of those $n + v$ boxes in a row. In each box preceding the first marked box we put a factor x_1, in the unmarked boxes between the marks no. 1 and no. 2 the factor x_2, in the boxes between the marks no. 2 and no. 3 the factor $x_3, \ldots$, in the boxes between the marks no. $(v - 1)$ and no. v the factor x_v, in each box following the last marked box the factor 1. Thus to any choice of v boxes out of those $n + v$ corresponds a product of the form $x_1^{m_1} x_2^{m_2} x_3^{m_3} \cdots x_v^{m_v}$ with $m_1 + m_2 + m_3 + \cdots + m_v \leq n$ and so a term of the polynomial: This is an intuitive argument.

Cf. ex. 3.40.1.

6.13.1. (Stanford 1949.) The problem is: Find a positive integer x such that all digits of x^2 are 1.

(1) *Keep only a part* (a small part) *of the condition:* The last digit of x^2 is 1. As the last digit of x^2 depends only on the last digit of x it is enough to consider

the one-digit numbers; their squares are

$$0, \quad 1, \quad 4, \quad 9, \quad 16, \quad 25, \quad 36, \quad 49, \quad 64, \quad 81$$

respectively. [Observe that x^2 and $(10 - x)^2$ have the same last digit.] Thus only numbers ending in 1 or 9 are eligible.

(2) *Keep a part* (the next larger part) *of the condition:* The last two digits of x^2 are 11. Enough to consider two-digit numbers, of the two numbers x and 100-x just one, and so, by virtue of (1), finally just the ten numbers 01, 11, 21, ... 91. None of them has a square with the ending 11: This proves the assertion.

Moral: It may be advantageous to transform a "problem to prove" into a "problem to find."

6.17.1. [Stanford 1965.] It is understood that for "age" we admit only a whole number of years. Here is a complete list of the decompositions of 72 into three positive integral factors each followed by the sum of the three factors:

$1 \cdot 1 \cdot 72$	74	$2 \cdot 2 \cdot 18$	22
$1 \cdot 2 \cdot 36$	39	$2 \cdot 3 \cdot 12$	17
$1 \cdot 3 \cdot 24$	28	$2 \cdot 4 \cdot 9$	15
$1 \cdot 4 \cdot 18$	23	$2 \cdot 6 \cdot 6$	**14**
$1 \cdot 6 \cdot 12$	19	$3 \cdot 3 \cdot 8$	**14**
$1 \cdot 8 \cdot 9$	18	$3 \cdot 4 \cdot 6$	13

The only sum of three factors that occurs more than once is emphasized by bold print. The remark about *the* oldest boy yields a distinction between two otherwise equally eligible cases: The boys are 8,3, and 3 years old.

7.2.1. See ex. 3.91. The points represent the quantities

$$C_6 \qquad I_6$$
$$C_{12}$$
$$\qquad I_{12}$$
$$C_{24}$$
$$\qquad I_{24}$$
$$C_{48}$$
$$\qquad I_{48}$$
$$C_{96}$$
$$\qquad I_{96}$$

9.11.1. (*Cf. American Mathematical Monthly,* vol. 56, 1949, pp. 423–424.) Choose in advance

$$\min (a_n, b_n) = \frac{1}{n^2}$$

(Any other convergent series with decreasing positive terms would do just as well.) The series

$$(1) + \left(\frac{1}{4}\right) + \left(\frac{1}{4} + \frac{1}{4} + \frac{1}{4} + \frac{1}{4}\right) + \left(\frac{1}{49} + \frac{1}{64} + \cdots + \frac{1}{42^2}\right) + \left(\frac{1}{42^2} + \cdots\right.$$

$$(1) + (1) + \left(\frac{1}{9} + \frac{1}{16} + \frac{1}{25} + \frac{1}{36}\right) + \left(\frac{1}{36} + \frac{1}{36} + \cdots + \frac{1}{36}\right) + \left(\frac{1}{43^2} + \cdots\right.$$

are constructed in corresponding "segments" indicated by parentheses. In one segment each term equals the minimum chosen in advance; in the corresponding segment of the other series all terms are equal to each other and to the last term of the preceding segment, and their sum is 1. These two kinds of segments alternate in both series.

These series do not yet fulfill quite the condition stated: The terms are decreasing in the sense "$\geq$" and not yet in the sense "$>$." Yet there is an easy modification: In segments where the consecutive terms are all equal, subtract from them consecutive terms of an arithmetic progression with sufficiently small positive initial term and difference, especially with sum $< \frac{1}{2}$.

12.2.1. (Stanford 1952.) *What is the hypothesis?*

$$a < \frac{b + c}{2}$$

What is the conclusion?

$$\alpha < \frac{\beta + \gamma}{2}$$

which is equivalent to

$$2\alpha < \pi - \alpha$$
$$\alpha < \pi/3$$

Relevant knowledge:

$$a^2 = b^2 + c^2 - 2bc \cos \alpha$$

$$\cos \alpha = \frac{b^2 + c^2 - a^2}{2bc}$$

$$> \frac{b^2 + c^2 - (b + c)^2/4}{2bc}$$

$$= \frac{3(b^2 + c^2)}{8bc} - \frac{1}{4}$$

$$\geq \frac{6bc}{8bc} - \frac{1}{4} = \frac{1}{2} = \cos \frac{\pi}{3}$$

which proves the assertion.

12.5.1. (Julius G. Baron; see *Mathematics Magazine* **39**, 1966, p. 134 and p. 112.)

The problem is substantially: Find the area A of a quadrilateral which is both inscribed and circumscribed, given its sides a,b,c, and d.

A closely related problem was raised in India twelve centuries ago. You stand a good chance to remember it if you have ever heard of it and ask yourself earnestly: *Do you know a related problem? Do you know a problem with the same kind of unknown?*

In fact, that related problem has the same unknown and the same data as the proposed; it has even one-half of the most striking clause of the proposed condition; it is: *Find the area* A *of an inscribed quadrilateral being given its sides* a,b,c, *and* d. Its solution is (MPR, vol. 1, pp. 251–252, ex. 41)

$$A^2 = (p - a)(p - b)(p - c)(p - d)$$

where $2p = a + b + c + d$.

In possession of this information, it is sufficient to observe: If the quadrilateral is circumscribed and side a is opposite to c, side b to d, then

$$a + c = b + d = p$$

12.9.1. Set

$$b + c - a = 2A \qquad c + a - b = 2B \qquad a + b - c = 2C$$

Then

$$\frac{1}{x^2} = \frac{A}{xyz} \qquad \frac{1}{y^2} = \frac{B}{xyz} \qquad \frac{1}{z^2} = \frac{C}{xyz}$$

By multiplying, we obtain

$$xyz = ABC$$

and so

$$x^2 = BC \qquad y^2 = CA \qquad z^2 = AB$$

For a full treatment there remains the discussion of the case when one or more of the unknowns x, y, and z vanish.

12.9.2. The usual definition of the ellipse mentions the foci. The consideration of this definition may lead to the question "Where are the foci?" and so eventually to guessing a *stronger proposition which is easier to prove:* Under the given assumptions concerning f, f', and v, the locus of V is an ellipse with foci F and F'. In fact, the definition of the ellipse is satisfied. As the figure easily shows

$$FV + F'V = r + r'$$

where r is the radius of f and r' that of f'.

12.9.3. (1) Concentric circle with radius $(r + r')/2$. (2) Parabola with directrix f and focus F'. (3) Hyperbola with foci F and F'. (4) Perpendicular bisector of the segment with endpoints $f = F$ and $f' = F'$.

Some other cases:

(5) Limiting case of ex. 12.9.2, or of (3): f' is a point on f, the locus is a straight line.

(6) Particular case of (3): $r = r'$, the locus is a straight line.

There are more questions.

About (3): What is the direction of the asymptotes, where is their point of intersection? Direction and location must be determined by the fixed circles f and f', but how? And why?

About (5): Is the whole straight line a limiting case of an ellipse? Or that of a hyperbola? Or is some part of the line this and the other that?

And so on.

To the teacher: The case in which f' is a point is beautifully presented by four films in the series *Animated Geometry* by J. L. Nicolet.

14.1.1. If 400 consecutive Gregorian years precisely agreed with 400 astronomical years, the length of one astronomical year would be

$$\frac{97 \cdot 366 + 303 \cdot 365}{400} = 365 + \frac{97}{400}$$

days, that is, 365 days, 5 hours, 49 minutes, and 12 seconds, which is only 26 seconds longer than the length derived from astronomical observations. The discrepancy is small, but it would amount to a day in 3323 years.

15.2.1. (*Cf.* Stanford 1964.)   (1) The segments into which d divides c are proportional to the adjacent sides.   Therefore

$$\left(\frac{ac}{a+b}\right)^2 = a^2 + d^2 - 2ad \cos \frac{\gamma}{2}$$

$$\left(\frac{bc}{a+b}\right)^2 = b^2 + d^2 - 2bd \cos \frac{\gamma}{2}$$

Straightforward elimination of γ yields

$$d^2 = \frac{ab[(a+b)^2 - c^2]}{(a+b)^2}$$

(2) If $a = b = c$, then $d^2 = 3a^2/4$.   If $a^2 + b^2 = c^2$, then

$$\frac{d}{a\sqrt{2}} = \frac{b}{a+b}$$

and this proportion is visible from easily constructed similar triangles.
   If $a = b$, then $d^2 = a^2 - (c/2)^2$.
   If $a + b = c$, then $d = 0$.

15.52.1. The formulation of the problem leaves considerable latitude and it does so intentionally: "real" problems may be rather indeterminate at the start. I detail some points deserving consideration.

   (1) The points of each set fill the interior and the surface of a *polyhedron*, see Figs. AS1, AS2, AS3.

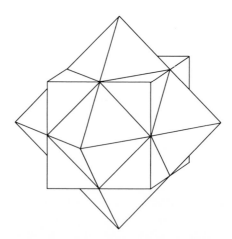

Fig. AS1.   See C and O, imagine I and H.

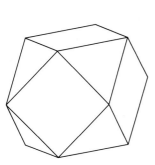

Fig. AS2. The intersection I.

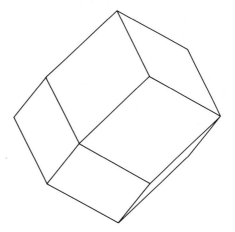

Fig. AS3. The convex hull H.

C is a cube, its faces are squares.

O is a regular octahedron, its faces are equilateral triangles (*Cf*. ex. 5.5).

I is the *intersection* of C and O; it is called the *cuboctahedron*. It has 14 faces: 6 faces are squares each cut out from a face of C, and 8 faces are equilateral triangles each cut out from a face of O.

H contains both C and O; it is, in fact, the smallest convex set containing both, their *convex hull*. Its faces are rhombi, it is called the *rhombic dodecahedron*.

We pass from C to I by slicing off from C 8 congruent tetrahedra.

We pass from C to H by adding to C 6 congruent pyramids.

(2) We list the vertices of our four polyhedra:

C	$(\pm 1, \pm 1, \pm 1)$		
O	$(\pm 2, 0, 0,)$	$(0, \pm 2, 0)$	$(0, 0, \pm 2)$
I	$(0, \pm 1, \pm 1)$	$(\pm 1, 0, \pm 1)$	$(\pm 1, \pm 1, 0)$
H	vertices of both C and O.		

(3) The following table exhibits F, V, and E, the number of faces, vertices, and edges, for each polyhedron.

	F	V	E
C	6	8	12
O	8	6	12
I	$6 + 8$	12	24
H	12	$8 + 6$	24

(4) C, O, and I are inscribed in H. Eight among the 14 vertices of H are the vertices of C, the remaining 6 the vertices of O, the centers of the 12 faces of H are the vertices of I.

Each edge of C is coupled with an edge of O: they bisect each other, they are the diagonals of the same face of H, their point of intersection is a vertex of I.

(5) All four polyhedra have the same kind of symmetry; we describe it in considering the most familiar figure C.

There are *planes of symmetry* of two different kinds in the cube:

Three are parallel, and midway between, a pair of opposite faces of the cube. Six pass through a pair of opposite edges.

All 9 planes of symmetry pass through the center of the cube and cut it into 48 congruent tetrahedra.

There are *axes of symmetry* of three different kinds. (*cf.* HSI, ex. 8, pp. 235, 239, 244–245) in the cube:

Six connect the midpoints of a pair of opposite edges, each one is the intersection of 2 planes of symmetry.

Four connect a pair of opposite vertices, each one is the intersection of 3 planes of symmetry.

Three connect the midpoints of a pair of opposite faces, each one is the intersection of 4 planes of symmetry.

All 13 axes of symmetry pass through the center of the cube. If the axis is an intersection of n planes of symmetry, then the cube rotated about the axis through an angle $360°/n$ coincides with itself.

(6) The two patterns of ex. 2.50.1 involve C and H, respectively. In the first pattern, each sphere is inscribed in a cube and all these cubes fill the whole space without gaps and overlapping. In the second pattern, each sphere is inscribed in a rhombic dodecahedron and all these rhombic dodecahedra fill the whole space without gaps and overlapping.

(7) To compute V [not the same notation as under (3)!] for I and H, we can advantageously start from C. If the polyhedron is circumscribed about a sphere there is a connection between V and S.

C	$S = 24$	$V = \dfrac{1}{3} 1 \cdot S = 8$
O	$S = 16\sqrt{3}$	$V = \dfrac{1}{3} \dfrac{2}{\sqrt{3}} S = \dfrac{32}{3}$
I	$S = 12 + 4\sqrt{3}$	$V = \dfrac{20}{3}$
H	$S = 24\sqrt{2}$	$V = \dfrac{1}{3}\sqrt{2}\,S = 16$

(8) The example may serve to introduce several general concepts, for instance, systems of linear inequalities, convex hull, symmetry in space, etc.

Some more particular questions: Are there other pairs of polyhedra so coupled as C and O? Other space filling polyhedra? Etc.

15.53.1. Let a, b, and D be positive integers, D not a perfect square, so linked that

$$a^2 - b^2 D = 1$$

$a = 2, b = 1, D = 3$ is an example. Let n be a positive integer; then there are integers A and B such that

$$(a - b\sqrt{D})^n = A - B\sqrt{D}$$

There follows
$$(a + b\sqrt{D})^n = A + B\sqrt{D}$$
$$\begin{aligned} A^2 - B^2D &= (a + b\sqrt{D})^n(a - b\sqrt{D})^n \\ &= (a^2 - b^2D)^n \\ &= 1 \end{aligned}$$
and
$$\begin{aligned} (a - b\sqrt{D})^n &= \sqrt{A^2} - \sqrt{B^2D} \\ &= \sqrt{A^2} - \sqrt{A^2 - 1} \end{aligned}$$

Just a little change is needed to similarly generalize ex. 15.53, or to merge it with the present problem into a common generalization.

INDEX

This index contains references to both volumes of the present work and, moreover, references to selected parallel passages of my related works. **Heavy print** is used for distinguishing from each other the five volumes in question:

1 for volume 1 of *Mathematical Discovery*
2 for volume 2 of *Mathematical Discovery* (the present volume)
H for *How to Solve It,* second edition
I for volume 1 of *Mathematics and Plausible Reasoning* (*Induction and Analogy in Mathematics*)
II for volume 2 of *Mathematics and Plausible Reasoning* (*Patterns of Plausible Inference*).

The numbers in ordinary type, following a sign in heavy type, refer to pages.
Italics are used to emphasize certain sentences whose role is discussed in chapter 12.
The index does not (and could not reasonably) give exhaustive references to topics treated virtually throughout all five volumes such as "Analogy" "Generalization" "Guess" "Induction" "Unknown" etc.